facts101
Just The

Textbook Key Facts

Textbook Outlines, Highlights, and Practice Quizzes

Business Communication Today

by Courtland Bovee, 11th Edition

All "Just the Facts101" Material Written or Prepared by Cram101 Publishing

---

Title Page

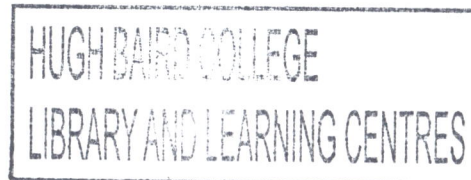

Visit Cram101.com for full Practice Exams

T68335

cram101
LEARNING SYSTEM

"Just the Facts101" is a Cram101 publication and tool designed to give you all the facts from your textbooks. Visit Cram101.com for the full practice test for each of your chapters for virtually any of your textbooks.

Cram101 has built custom study tools specific to your textbook. We provide all of the factual testable information and unlike traditional study guides, we will never send you back to your textbook for more information.

YOU WILL NEVER HAVE TO HIGHLIGHT A BOOK AGAIN!

Cram101 StudyGuides

All of the information in this StudyGuide is written specifically for your textbook. We include the key terms, places, people, and concepts... the information you can expect on your next exam!

Want to take a practice test?

Throughout each chapter of this StudyGuide you will find links to cram101.com where you can select specific chapters to take a complete test on, or you can subscribe and get practice tests for up to 12 of your textbooks, along with other exclusive cram101.com tools like problem solving labs and reference libraries.

Cram101.com

Only cram101.com gives you the outlines, highlights, and PRACTICE TESTS specific to your textbook. Cram101.com is an online application where you'll discover study tools designed to make the most of your limited study time.

By purchasing this book, you get 50% off the normal monthly subscription fee!. Just enter the promotional code **'DK73DW21191'** on the Cram101.com registration screen.

www.Cram101.com

Learning System

facts101

Business Communication Today
Courtland Bovee, 11th

CONTENTS

_____ Business communication

_____ Social media

_____ Statement

_____ User-generated content

_____ Search Engine

_____ Social network

_____ Labor force

_____ Stakeholder

_____ Globalization

_____ Interactivity

_____ Intercultural communication

_____ Job interview

_____ Knowledge worker

_____ Entrepreneurial culture

_____ Marketing Intelligence

_____ Organizational structure

_____ Viral marketing

_____ Cultural identity

_____ Upward communication

FreshBooks

Audience analysis

Emotional intelligence

Public speaking

Receiver

Crowdsourcing

Selective perception

Get Satisfaction

Long-term memory

Online shopping

Sensory memory

Short-term memory

YouTube

Information overload

Perspective

Sexual harassment

Collaborative workspace

Unemployment

Workspace

CHAPTER OUTLINE: KEY TERMS, PEOPLE, PLACES, CONCEPTS

_____ | Survey research

_____ | Social bookmarking

_____ | Social tagging

_____ | Supply chain

_____ | Supply chain management

_____ | Microblogging

_____ | Podcasts

_____ | Information technology

_____ | Productivity

_____ | Distortion

_____ | Comparative advertising

_____ | Defamation

_____ | Discovery

_____ | Financial Reporting

_____ | Information product

_____ | Intellectual property

_____ | Job description

_____ | Proposal

_____ | Social networking

Chapter 1. Achieving Success Through Effective Business Communication

Business communication	Business Communication: communication used to promote a product, service, or organization; relay information within the business; or deal with legal and similar issues. It is also a means of relaying between a supply chain, for example the consumer and manufacturer. Business Communication is known simply as 'communications'.
Social media	Social media includes web- and mobile-based technologies which are used to turn communication into interactive dialogue among organizations, communities, and individuals. Andreas Kaplan and Michael Haenlein define social media as 'a group of Internet-based applications that build on the ideological and technological foundations of Web 2.0, and that allow the creation and exchange of user-generated content.' When the technologies are in place, social media is ubiquitously accessible, and enabled by scalable communication techniques. Social media technologies take on many different forms including magazines, Internet forums, weblogs, social blogs, microblogging, wikis, social networks, podcasts, photographs or pictures, video, rating and social bookmarking.
Statement	In logic a statement is either (a) a meaningful declarative sentence that is either true or false, or (b) what is asserted or made by the use of a declarative sentence. In the latter case, a statement is distinct from a sentence in that a sentence is only one formulation of a statement, whereas there may be many other formulations expressing the same statement. Philosopher of language, Peter Strawson advocated the use of the term 'statement' in sense (b) in preference to proposition.
User-generated content	User-generated content covers a range of media content available in a range of modern communications technologies. It entered mainstream usage during 2005, having arisen in web publishing and new media content production circles. Its use for a wide range of applications, including problem processing, news, gossip and research, reflects the expansion of media production through new technologies that are accessible and affordable to the general public.
Search Engine	Search Engine was a weekly Canadian radio show that aired on CBC Radio One, then a dedicated podcast available on CBC.ca, and now a podcast on TVOntario's website, tvo.org. It is hosted by Jesse Brown, who also co-produces the show with Geoff Siskind and Andrew Parker. Cory Doctorow, novelist and editor of Boing Boing, is also a regular contributor.
Social network	A social network is a social structure made up of a set of actors (such as individuals or organizations) and the dyadic ties between these actors. The social network perspective provides a clear way of analyzing the structure of whole social entities.

Chapter 1. Achieving Success Through Effective Business Communication

Labor force	Normally, the labor force of a country consists of everyone of working age (typically above a certain age (around 14 to 16) and below retirement (around 65) who are participating workers, that is people actively employed or seeking employment. People not counted include students, retired people, stay-at-home parents, people in prisons or similar institutions, people employed in jobs or professions with unreported income, as well as discouraged workers who cannot find work. In the United States, the unemployment rate is estimated by a household survey called the Current Population Survey, conducted monthly by the Federal Bureau of Labor Statistics.
Stakeholder	A corporate stakeholder is a party that can affect or be affected by the actions of the business as a whole. The stakeholder concept was first used in a 1963 internal memorandum at the Stanford Research Institute. It defined stakeholders as 'those groups without whose support the organization would cease to exist.' The theory was later developed and championed by R. Edward Freeman in the 1980s.
Globalization	Globalization is a common term for processes of international integration arising from increasing human connectivity and interchange of worldviews, products, ideas, and other aspects of culture. In particular, advances in transportation and telecommunications infrastructure, including the rise of the Internet, represent major driving factors in globalization and precipitate further interdependence of economic and cultural activities. Though several scholars situate the origins of globalization in modernity, others map its history long before the European age of discovery and voyages to the New World.
Interactivity	In the fields of information science, communication, and industrial design, there is debate over the meaning of interactivity. In the 'contingency view' of interactivity, there are three levels:•Noninteractive, when a message is not related to previous messages;•Reactive, when a message is related only to one immediately previous message; and•Interactive, when a message is related to a number of previous messages and to the relationship between them. Human to human communication Human communication is the basic example of interactive communication which involves two different processes; human to human interactivity and human to computer interactivity. Human-Human interactivity is the communication between people.
Intercultural communication	Intercultural communication is a form of global communication. It is used to describe the wide range of communication problems that naturally appear within an organization made up of individuals from different religious, social, ethnic, and educational backgrounds. Intercultural communication is sometimes used synonymously with cross-cultural communication.

Chapter 1. Achieving Success Through Effective Business Communication

Job interview	A job interview is a process in which a potential employee is evaluated by an employer for prospective employment in their company, organization, or firm. During this process, the employer hopes to determine whether or not the applicant is suitable for the role. A job interview typically precedes the hiring decision, and is used to evaluate the candidate.
Knowledge worker	Knowledge workers are workers whose main capital is knowledge. Typical examples may include software engineers, architects, engineers, scientists and lawyers, because they 'think for a living'. What differentiates knowledge work from other forms of work is its primary task of 'non-routine' problem solving that requires a combination of convergent, divergent, and creative thinking (Reinhardt et al., 2011).
Entrepreneurial culture	Entrepreneurial culture, is a form of ideal, which is based on the value system of an enterprise and closely related to the management philosophy as well as the management behaviour of the enterprise. It is where the kernel of business management lies. In the narrow sense, it refers to some fundamental spirit and agglomerating force that come into being in the production and management practices of an enterprise, as well as the common values and norms of behavior shared by the whole staff.
Marketing Intelligence	Marketing Intelligence is the information relevant to a company's markets, gathered and analyzed specifically for the purpose of accurate and confident decision-making in determining market opportunity, market penetration strategy, and market development metrics. Marketing intelligence is necessary when entering a foreign market. Marketing Intelligence is not the same as Market Intelligence (MARKINT).
Organizational structure	An organizational structure consists of activities such as task allocation, coordination and supervision, which are directed towards the achievement of organizational aims. It can also be considered as the viewing glass or perspective through which individuals see their organization and its environment. Organizations are a variant of clustered entities.
Viral marketing	Viral marketing, viral advertising, or marketing buzz are buzzwords referring to marketing techniques that use pre-existing social networks to produce increases in brand awareness or to achieve other marketing objectives (such as product sales) through self-replicating viral processes, analogous to the spread of viruses or computer viruses (cf. internet memes and memetics). It can be delivered by word of mouth or enhanced by the network effects of the Internet.

Cultural identity	Cultural identity is the identity of a group or culture, or of an individual as far as one is influenced by one's belonging to a group or culture. Cultural identity is similar to and has overlaps with, but is not synonymous with, identity politics. Description
	Various modern cultural studies and social theories have investigated cultural identity.
Upward communication	Upward Communication is the process of information flowing from the lower levels of a hierarchy to the upper levels. This type of communication is becoming more and more popular in organizations as traditional forms of communication are becoming less popular. The more traditional organization types such as a hierarchy, places people into separate ranks.
FreshBooks	FreshBooks is an online invoicing software as a service for freelancers, small businesses, agencies, and professionals. It is produced by the software company 2ndSite Inc. which is located in Toronto, Ontario, Canada.
Audience analysis	Audience analysis is a task that is often performed by technical writers in a project's early stages. It consists of assessing the audience to make sure the information provided to them is at the appropriate level. The audience is often referred to as the end-user, and all communications need to be targeted towards the defined audience.
Emotional intelligence	Emotional intelligence is the ability to identify, assess, and control the emotions of oneself, of others, and of groups. Various models and definitions have been proposed of which the ability and trait EI models are the most widely accepted in the scientific literature. Ability EI is usually measured using maximum performance tests and has stronger relationships with traditional intelligence, whereas trait EI is usually measured using self-report questionnaires and has stronger relationships with personality.
Public speaking	Public speaking is the process of speaking to a group of people in a structured, deliberate manner intended to inform, influence, or entertain the listeners. It is closely allied to 'presenting', although the latter has more of a commercial advertisement.
	In public speaking, as in any form of communication, there are five basic elements, often expressed as 'who is saying what to whom using what medium with what effects?' The purpose of public speaking can range from simply transmitting information, to motivating people to act, to simply telling a story.
Receiver	In modulated ultrasound terminology, a receiver is a device that receives a modulated ultrasound signal and decodes it for use as sound, navigational-position information, etc. Its function is somewhat like that of a radio receiver
Crowdsourcing	Crowdsourcing is a process that involves outsourcing tasks to a distributed group of people.

	This process can occur both online and offline. The difference between crowdsourcing and ordinary outsourcing is that a task or problem is outsourced to an undefined public rather than a specific body, such as paid employees.
Selective perception	Selective Perception is the process by which individuals perceive what they want to in media messages and disregard the rest. It is a broad term to identify the behavior all people exhibit to tend to 'see things' based on their particular frame of reference. Selective perception may refer to any number of cognitive biases in psychology related to the way expectations affect perception.
Get Satisfaction	Get Satisfaction is a customer community software platform for technical support based in San Francisco, California, United States. It was founded on January 31, 2007 by several people, including Lane Becker, Amy Muller, Thor Muller, and Jonathan Grubb. It publicly launched in September 2007. The idea for the service originated from Valleyschwag as a side project.
Long-term memory	Long-term memory is memory in which associations among items are stored, as part of the theory of a dual-store memory model. According to the theory, long-term memory differs structurally and functionally from working memory or short-term memory, which ostensibly stores items for only around 20-30 seconds and can be recalled easily. This differs from the theory of the single-store retrieved context model that has no differentiation between short-term and long-term memory.
Online shopping	Online shopping is a form of electronic commerce whereby consumers directly buy goods or services from a seller over the Internet without an intermediary service. An online shop, eshop, e-store, Internet shop, webshop, webstore, online store, or virtual store evokes the physical analogy of buying products or services at a bricks-and-mortar retailer or shopping centre. The process is called business-to-consumer (B2C) online shopping.
Sensory memory	During every moment of an organism's life, sensory information is being taken in by sensory receptors and processed by the nervous system. Humans have five main senses: sight, hearing, taste, smell, touch. Sensory memory allows individuals to retain impressions of sensory information after the original stimulus has ceased.
Short-term memory	Short-term memory is the capacity for holding a small amount of information in mind in an active, readily available state for a short period of time. The duration of short-term memory is believed to be in the order of seconds. A commonly cited capacity is 7 ± 2 elements.
YouTube	YouTube is a video-sharing website, created by three former PayPal employees in February 2005, on which users can upload, view and share videos. The company is based in San Bruno, California, and uses Adobe Flash Video and HTML5 technology to display a wide variety of user-generated video content, including movie clips, TV clips, and music videos, as well as amateur content such as video blogging, short original videos, and educational videos.

Information overload	'Information overload' is a term popularized by Alvin Toffler in his bestselling 1970 book Future Shock. It refers to the difficulty a person can have understanding an issue and making decisions that can be caused by the presence of too much information. The term itself is mentioned in a 1964 book by Bertram Gross, The Managing of Organizations.
Perspective	Perspective in the graphic arts, such as drawing, is an approximate representation, on a flat surface (such as paper), of an image as it is seen by the eye. The two most characteristic features of perspective are that objects are drawn:•Smaller as their distance from the observer increases•Foreshortened: the size of an object's dimensions along the line of sight are relatively shorter than dimensions across the line of sightOverview Linear perspective always works by representing the light that passes from a scene through an imaginary rectangle (the painting), to the viewer's eye. It is similar to a viewer looking through a window and painting what is seen directly onto the windowpane.
Sexual harassment	Sexual harassment is intimidation, bullying or coercion of a sexual nature, or the unwelcome or inappropriate promise of rewards in exchange for sexual favors. In most modern legal contexts sexual harassment is illegal. As defined by EEOC, 'It is unlawful to harass a person (an applicant or employee) because of that person's sex.' Harassment can include 'sexual harassment' or unwelcome sexual advances, requests for sexual favors, and other verbal or physical harassment of a sexual nature.
Collaborative workspace	A collaborative workspace is an inter-connected environment in which all the participants in dispersed locations can access and interact with each other just as inside a single entity. The environment may be supported by electronic communications and groupware which enable participants to overcome space and time differentials. These are typically enabled by a shared mental model, common information, and a shared understanding by all of the participants regardless of physical location.
Unemployment	Unemployment , occurs when people are without jobs and they have actively sought work within the past four weeks. The unemployment rate is a measure of the prevalence of unemployment and it is calculated as a percentage by dividing the number of unemployed individuals by all individuals currently in the labor force. During periods of recession, an economy usually experiences a relatively high unemployment rate.
Workspace	Workspace is a term used in various branches of engineering and economic development. Workspace refers to small premises provided, often by local authorities or economic development agencies, to help new businesses to establish themselves.

Chapter 1. Achieving Success Through Effective Business Communication

Survey research	Survey research involves utilizing interviews or questionnaires to obtain quantitative information in fields such as marketing, politics, and social science. Utilizing surveys is considered to be an efficient way of collecting data from a large number of respondents, accurately representing a whole population. Surveys also have the benefit of providing data that is relatively free from errors.
Social bookmarking	Social bookmarking is a method for Internet users to organize, store, manage and search for bookmarks of resources online. Many online bookmark management services have launched since 1996; Delicious, founded in 2003, popularized the terms 'social bookmarking' and 'tagging'. Tagging is a significant feature of social bookmarking systems, enabling users to organize their bookmarks in flexible ways and develop shared vocabularies known as folksonomies.
Social tagging	A folksonomy is a system of classification derived from the practice and method of collaboratively creating and managing tags to annotate and categorize content; this practice is also known as collaborative tagging, social classification, social indexing, and social tagging. Folksonomy, a term coined by Thomas Vander Wal, is a portmanteau of folk and taxonomy. Folksonomies became popular on the Web around 2004 as part of social software applications such as social bookmarking and photograph annotation.
Supply chain	A supply chain is a system of organizations, people, technology, activities, information and resources involved in moving a product or service from supplier to customer. Supply chain activities transform natural resources, raw materials and components into a finished product that is delivered to the end customer. In sophisticated supply chain systems, used products may re-enter the supply chain at any point where residual value is recyclable.
Supply chain management	Supply chain management is the management of a network of interconnected businesses involved in the provision of product and service packages required by the end customers in a supply chain. Supply chain management spans all movement and storage of raw materials, work-in-process inventory, and finished goods from point of origin to point of consumption. Another definition is provided by the APICS Dictionary when it defines SCM as the 'design, planning, execution, control, and monitoring of supply chain activities with the objective of creating net value, building a competitive infrastructure, leveraging worldwide logistics, synchronizing supply with demand and measuring performance globally.'Origin of the term and definitions The term 'supply chain management' entered the public domain when Keith Oliver, a consultant at Booz Allen Hamilton, used it in an interview for the Financial Times in 1982.

Microblogging	Microblogging is a broadcast medium in the form of blogging. A microblog differs from a traditional blog in that its content is typically smaller in both actual and aggregate file size. Microblogs 'allow users to exchange small elements of content such as short sentences, individual images, or video links'.
Podcasts	A podcast is a type of digital media consisting of an episodic series of audio radio, video, PDF, or ePub files subscribed to and downloaded through web syndication or streamed online to a computer or mobile device. The word is a neologism derived from 'broadcast' and 'pod' from the success of the iPod, as podcasts are often listened to on portable media players. In the context of Apple devices, the term 'Podcasts' refers to the audio and video version of podcasts, whereas the textual version of podcasts are classified under the app known as Newsstand.
Information technology	Information technology is a branch of engineering dealing with the use of computers and telecommunications equipment to store, retrieve, transmit and manipulate data. The term is commonly used as a synonym for computers and computer networks, but it also encompasses other information distribution technologies such as television and telephones. Some of the modern and emerging fields of Information technology are next generation web technologies, bioinformatics, cloud computing, global information systems, large-scale knowledge bases, etc.
Productivity	Productivity is a measure of the efficiency of production. Productivity is a ratio of production output to what is required to produce it (inputs). The measure of productivity is defined as a total output per one unit of a total input.
Distortion	A distortion is departure from the allocation of economic resources from the state in which each agent maximizes her own welfare.. A proportional wage-income tax, for instance, is distortionary, whereas a Lump-sum tax is not. In a competitive equilibrium, a proportional wage income tax discourages work.
Comparative advertising	Comparative advertising is an advertisement in which a particular product, or service, specifically mentions a competitor by name for the express purpose of showing why the competitor is inferior to the product naming it. Comparative advertising, also known as 'knocking copy', is loosely defined as advertising where 'the advertised brand is explicitly compared with one or more competing brands and the comparison is oblivious to the audience.' This should not be confused with parody advertisements, where a fictional product is being advertised for the purpose of poking fun at the particular advertisement, nor should it be confused with the use of a coined brand name for the purpose of comparing the product without actually naming an actual competitor. ('Wikipedia tastes better and is less filling than the Encyclopedia Galactica.')

Chapter 1. Achieving Success Through Effective Business Communication

Defamation	Defamation--also called calumny, vilification, traducement, slander (for transitory statements), and libel (for written, broadcast, or otherwise published words)--is the communication of a statement that makes a claim, expressly stated or implied to be factual, that may give an individual, business, product, group, government, or nation a negative image. This can be also any disparaging statement made by one person about another, which is communicated or published, whether true or false, depending on legal state. In Common Law it is usually a requirement that this claim be false and that the publication is communicated to someone other than the person defamed (the claimant).
Discovery	Discovery is the act of detecting something new, or something 'old' that had been unknown. With reference to science and academic disciplines, discovery is the observation of new phenomena, new actions, or new events and providing new reasoning to explain the knowledge gathered through such observations with previously acquired knowledge from abstract thought and everyday experiences. Visual discoveries are often called sightings.
Financial Reporting	Financial reporting is the process of preparing and distributing financial information to users of such information in various forms. The most common format of formal financial reporting are financial statements. Financial statements are prepared in accordance with rigorously applied standards defined by professional accounting bodies developed according to the legal and professional framework of a specific locale.
Information product	An Information Product is any final product in the form of information that a person needs to have. This Information Product consists of several Information Element, which are located in the organizational value chain. To illustrate the concept of an IP, an example is shown of a bottleneck analysis in HR (by J. Willems 2008).
Intellectual property	Intellectual property is a term referring to a number of distinct types of expressions for which a set of rights are recognized under the corresponding fields of law. Under intellectual property law, owners are granted certain exclusive rights to various markets, machines, musical, literary, and artistic works; discoveries and inventions; and applications. Common types of intellectual property rights include copyrights, trademarks, patents, industrial design rights, and trade secrets in some jurisdictions.
Job description	A job description is a list that a person might use for general tasks, or functions, and responsibilities of a position. It may often include to whom the position reports, specifications such as the qualifications or skills needed by the person in the job, or a salary range. Job descriptions are usually narrative, but some may instead comprise a simple list of competencies; for instance, strategic human resource planning methodologies may be used to develop a competency architecture for an organization, from which job descriptions are built as a shortlist of competencies.

Proposal	A business proposal is a written offer from a seller to a prospective buyer. Business proposals are often a key step in the complex sales process--i.e., whenever a buyer considers more than price in a purchase. There are three distinct categories of business proposals:•formally solicited•informally solicited•unsolicited. Solicited proposals are written in response to published requirements, contained in a Request for Proposal (RFP), Request for Quotation (RFQ), Request for Information (RFI) or an Invitation For Bid (IFB).
Social networking	A social networking service is an online service, platform, or site that focuses on facilitating the building of social networks or social relations among people who, for example, share interests, activities, backgrounds, or real-life connections. A social network service consists of a representation of each user (often a profile), his/her social links, and a variety of additional services. Most social network services are web-based and provide means for users to interact over the Internet, such as e-mail and instant messaging.

1. _____, is a form of ideal, which is based on the value system of an enterprise and closely related to the management philosophy as well as the management behaviour of the enterprise. It is where the kernel of business management lies. In the narrow sense, it refers to some fundamental spirit and agglomerating force that come into being in the production and management practices of an enterprise, as well as the common values and norms of behavior shared by the whole staff.

   a. Entrepreneurial ecosystem
   b. Entrepreneurial finance
   c. Entrepreneurial Management Center
   d. Entrepreneurial culture

2. . _____ is the process of speaking to a group of people in a structured, deliberate manner intended to inform, influence, or entertain the listeners. It is closely allied to 'presenting', although the latter has more of a commercial advertisement.

   In _____, as in any form of communication, there are five basic elements, often expressed as 'who is saying what to whom using what medium with what effects?' The purpose of _____ can range from simply transmitting information, to motivating people to act, to simply telling a story.

a. Puppet ruler
b. Regional autonomy
c. Public speaking
d. Term of office

3. _____, viral advertising, or marketing buzz are buzzwords referring to marketing techniques that use pre-existing social networks to produce increases in brand awareness or to achieve other marketing objectives (such as product sales) through self-replicating viral processes, analogous to the spread of viruses or computer viruses (cf. internet memes and memetics). It can be delivered by word of mouth or enhanced by the network effects of the Internet.

a. Word-of-mouth marketing
b. Wordmark
c. Workbooks.com
d. Viral marketing

4. _____ in the graphic arts, such as drawing, is an approximate representation, on a flat surface (such as paper), of an image as it is seen by the eye. The two most characteristic features of _____ are that objects are drawn:•Smaller as their distance from the observer increases•Foreshortened: the size of an object's dimensions along the line of sight are relatively shorter than dimensions across the line of sightOverview

Linear _____ always works by representing the light that passes from a scene through an imaginary rectangle (the painting), to the viewer's eye. It is similar to a viewer looking through a window and painting what is seen directly onto the windowpane.

a. Physionotrace
b. Planimeter
c. Perspective
d. Plumbing drawing

5. _____ is the ability to identify, assess, and control the emotions of oneself, of others, and of groups. Various models and definitions have been proposed of which the ability and trait EI models are the most widely accepted in the scientific literature. Ability EI is usually measured using maximum performance tests and has stronger relationships with traditional intelligence, whereas trait EI is usually measured using self-report questionnaires and has stronger relationships with personality.

a. Edith of Wilton
b. Electronic performance support systems
c. Emotional intelligence
d. InterPress

1. d
2. c
3. d
4. c
5. c

You can take the complete Chapter Practice Test

for Chapter 1. Achieving Success Through Effective Business Communication
on all key terms, persons, places, and concepts.

Online 99 Cents

http://www.epub2.4.21191.1.cram101.com/

Use www.Cram101.com for all your study needs

including Cram101's online interactive problem solving labs in

chemistry, statistics, mathematics, and more.

CHAPTER OUTLINE: KEY TERMS, PEOPLE, PLACES, CONCEPTS

_____ | Law firm

_____ | Survey research

_____ | Cross-functional team

_____ | Social network

_____ | Cultural identity

_____ | Group dynamics

_____ | Norm

_____ | Public speaking

_____ | Development

_____ | Resignation

_____ | Collaborative workspace

_____ | Access control

_____ | Cloud computing

_____ | Content management

_____ | Content management system

_____ | Groupware

_____ | Management system

_____ | Negotiation

_____ | Rollback

Sexual harassment

Viral marketing

Virtual office

Web content management system

Workspace

Social networking

Virtual community

Virtual team

Productivity

Public opinion

Parliamentary procedure

Virtual business

Brainstorming

Virtual world

Active listening

Receiver

Long-term memory

Online shopping

Short-term memory

CHAPTER OUTLINE: KEY TERMS, PEOPLE, PLACES, CONCEPTS

	Nonverbal communication

	Statement

	Body language

	Nonverbal

	Personal space

	Topics

	Workforce management

	Appearance

	Business ethics

	Eye contact

	Small business

	Crowdsourcing

	Marketing Intelligence

Chapter 2. Mastering Team Skills and Interpersonal Communication

Law firm	A law firm is a business entity formed by one or more lawyers to engage in the practice of law. The primary service rendered by a law firm is to advise clients (individuals or corporations) about their legal rights and responsibilities, and to represent clients in civil or criminal cases, business transactions, and other matters in which legal advice and other assistance are sought. Law firms are organized in a variety of ways, depending on the jurisdiction in which the firm practices.
Survey research	Survey research involves utilizing interviews or questionnaires to obtain quantitative information in fields such as marketing, politics, and social science. Utilizing surveys is considered to be an efficient way of collecting data from a large number of respondents, accurately representing a whole population. Surveys also have the benefit of providing data that is relatively free from errors.
Cross-functional team	A cross-functional team is a group of people with different functional expertise working toward a common goal. It may include people from finance, marketing, operations, and human resources departments. Typically, it includes employees from all levels of an organization.
Social network	A social network is a social structure made up of a set of actors (such as individuals or organizations) and the dyadic ties between these actors. The social network perspective provides a clear way of analyzing the structure of whole social entities. The study of these structures uses social network analysis to identify local and global patterns, locate influential entities, and examine network dynamics.
Cultural identity	Cultural identity is the identity of a group or culture, or of an individual as far as one is influenced by one's belonging to a group or culture. Cultural identity is similar to and has overlaps with, but is not synonymous with, identity politics. Description Various modern cultural studies and social theories have investigated cultural identity.
Group dynamics	Group dynamics refers to a system of behaviors and psychological processes occurring within a social group (intragroup dynamics), or between social groups (intergroup dynamics). The study of group dynamics can be useful in understanding decision-making behavior, tracking the spread of diseases in society, creating effective therapy techniques, and following the emergence and popularity of new ideas and technologies. Group dynamics are at the core of understanding racism, sexism, and other forms of social prejudice and discrimination.
Norm	In linear algebra, functional analysis and related areas of mathematics, a norm is a function that assigns a strictly positive length or size to all vectors in a vector space, other than the zero vector (which has zero length assigned to it).

A seminorm, on the other hand, is allowed to assign zero length to some non-zero vectors (in addition to the zero vector).

A simple example is the 2-dimensional Euclidean space R^2 equipped with the Euclidean norm.

Public speaking	Public speaking is the process of speaking to a group of people in a structured, deliberate manner intended to inform, influence, or entertain the listeners. It is closely allied to 'presenting', although the latter has more of a commercial advertisement. In public speaking, as in any form of communication, there are five basic elements, often expressed as 'who is saying what to whom using what medium with what effects?' The purpose of public speaking can range from simply transmitting information, to motivating people to act, to simply telling a story.
Development	In classical differential geometry, development refers to the simple idea of rolling one smooth surface over another in Euclidean space. For example, the tangent plane to a surface (such as the sphere or the cylinder) at a point can be rolled around the surface to obtain the tangent-plane at other points. The tangential contact between the surfaces being rolled over one another provides a relation between points on the two surfaces.
Resignation	A resignation is the formal act of giving up or quitting one's office or position. It can also refer to the act of admitting defeat in a game like chess, indicated by the resigning player declaring 'I resign', turning his king on its side, extending his hand, or stopping the chess clock. A resignation can occur when a person holding a position gained by election or appointment steps down, but leaving a position upon the expiration of a term is not considered resignation.
Collaborative workspace	A collaborative workspace is an inter-connected environment in which all the participants in dispersed locations can access and interact with each other just as inside a single entity. The environment may be supported by electronic communications and groupware which enable participants to overcome space and time differentials. These are typically enabled by a shared mental model, common information, and a shared understanding by all of the participants regardless of physical location.
Access control	Access control refers to exerting control over who can interact with a resource. Often but not always, this involves an authority, who does the controlling. The resource can be a given building, group of buildings, or computer-based information system.

Chapter 2. Mastering Team Skills and Interpersonal Communication

Cloud computing	Cloud computing is the use of computing resources (hardware and software) that are delivered as a service over a network (typically the Internet). The name comes from the use of a cloud-shaped symbol as an abstraction for the complex infrastructure it contains in system diagrams. Cloud computing entrusts remote services with a user's data, software and computation.
Content management	Content management, is the set of processes and technologies that support the collection, managing, and publishing of information in any form or medium. In recent times this information is typically referred to as content or, to be precise, digital content. Digital content may take the form of text (such as electronic documents), multimedia files (such as audio or video files), or any other file type that follows a content lifecycle requiring management.
Content management system	A content management system is a computer system that allows publishing, editing, and modifying content as well as site maintenance from a central page. It provides a collection of procedures used to manage workflow in a collaborative environment. These procedures can be manual or computer-based.
Groupware	Collaborative software or groupware is computer software designed to help people involved in a common task achieve goals. One of the earliest definitions of collaborative software is 'intentional group processes plus software to support them.' The design intent of collaborative software is to transform the way documents and rich media are shared to enable more effective team collaboration. Collaboration, with respect to information technology, seems to have several definitions.
Management system	A management system is the framework of processes and procedures used to ensure that an organization can fulfill all tasks required to achieve its objectives. For instance, an environmental management system enables organizations to improve their environmental performance through a process of continuous improvement. An oversimplification is 'Plan, Do, Check, Act'.
Negotiation	Negotiation is a dialogue between two or more people or parties, intended to reach an understanding, resolve point of difference, or gain advantage in outcome of dialogue, to produce an agreement upon courses of action, to bargain for individual or collective advantage, to craft outcomes to satisfy various interests of two people/parties involved in negotiation process. Negotiation is a process where each party involved in negotiating tries to gain an advantage for themselves by the end of the process. Negotiation is intended to aim at compromise.
Rollback	In database technologies, a rollback is an operation which returns the database to some previous state. Rollbacks are important for database integrity, because they mean that the database can be restored to a clean copy even after erroneous operations are performed.

Sexual harassment	Sexual harassment is intimidation, bullying or coercion of a sexual nature, or the unwelcome or inappropriate promise of rewards in exchange for sexual favors. In most modern legal contexts sexual harassment is illegal. As defined by EEOC, 'It is unlawful to harass a person (an applicant or employee) because of that person's sex.' Harassment can include 'sexual harassment' or unwelcome sexual advances, requests for sexual favors, and other verbal or physical harassment of a sexual nature.
Viral marketing	Viral marketing, viral advertising, or marketing buzz are buzzwords referring to marketing techniques that use pre-existing social networks to produce increases in brand awareness or to achieve other marketing objectives (such as product sales) through self-replicating viral processes, analogous to the spread of viruses or computer viruses (cf. internet memes and memetics). It can be delivered by word of mouth or enhanced by the network effects of the Internet.
Virtual office	A virtual office is a combination of off-site live communication and address services that allow users to reduce traditional office costs while maintaining business professionalism. Frequently the term is confused with 'office business centers' or 'executive suites' which demand a conventional lease whereas a true virtual office does not require that expense. The virtual office idea came from the convergence of technological innovation and the Information Age.
Web content management system	A Web Content Management System is a software system that provides website authoring, collaboration, and administration tools designed to allow users with little knowledge of web programming languages or markup languages to create and manage website content with relative ease. A robust provides the foundation for collaboration, offering users the ability to manage documents and output for multiple author editing and participation. Most systems use a Content Repository or a database to store page content, metadata, and other information assets that might be needed by the system.
Workspace	Workspace is a term used in various branches of engineering and economic development. Workspace refers to small premises provided, often by local authorities or economic development agencies, to help new businesses to establish themselves. These typically provide not only physical space and utilities, but also administrative services and links to support and finance organisations, as well as peer support among the tenants.
Social networking	A social networking service is an online service, platform, or site that focuses on facilitating the building of social networks or social relations among people who, for example, share interests, activities, backgrounds, or real-life connections.

Chapter 2. Mastering Team Skills and Interpersonal Communication

	A social network service consists of a representation of each user (often a profile), his/her social links, and a variety of additional services. Most social network services are web-based and provide means for users to interact over the Internet, such as e-mail and instant messaging.
Virtual community	A virtual community is a social network of individuals who interact through specific social media, potentially crossing geographical and political boundaries in order to pursue mutual interests or goals. One of the most pervasive types of virtual community include social networking services, which consist of various online communities. The term virtual community is attributed to the book of the same title by Howard Rheingold, published in 1993. The book, which could be considered a social enquiry, putting the research in the social sciences, discussed his adventures on The WELL and onward into a range of computer-mediated communication and social groups, broadening it to information science.
Virtual team	A virtual team is a group of individuals who work across time, space and organizational boundaries with links strengthened by webs of communication technology' Ale Ebrahim, N., Ahmed, S. & Taha, Z. in their recent (2009) literature review paper, added two key issues to definition of a virtual team 'as small temporary groups of geographically, organizationally and/ or time dispersed knowledge workers who coordinate their work predominantly with electronic information and communication technologies in order to accomplish one or more organization tasks' Members of virtual teams communicate electronically and may never meet face-to-face. Virtual teams are made possible by a proliferation of fiber optic technology that has significantly increased the scope of off-site communication.
Productivity	Productivity is a measure of the efficiency of production. Productivity is a ratio of production output to what is required to produce it (inputs). The measure of productivity is defined as a total output per one unit of a total input.
Public opinion	Public opinion is the aggregate of individual attitudes or beliefs held by the adult population. Public opinion can also be defined as the complex collection of opinions of many different people and the sum of all their views. The principle approaches to the study of public opinion may be divided into 4 categories:•quantitative measurement of opinion distributions;•investigation of the internal relationships among the individual opinions that make up public opinion on an issue;•description or analysis of the public role of public opinion;•study both of the communication media that disseminate the ideas on which opinions are based and of the uses that propagandists and other manipulators make of these media.Concepts of public opinion Public opinion as a concept gained credence with the rise of 'public' in the eighteenth century.

Parliamentary procedure	Parliamentary procedure is the body of rules, ethics, and customs governing meetings and other operations of clubs, organizations, legislative bodies, and other deliberative assemblies. It is part of the common law originating primarily in the practices of the House of Commons of the Parliament of the United Kingdom, from which it derives its name. In the United States, parliamentary procedure is also referred to as parliamentary law, parliamentary practice, legislative procedure, or rules of order.
Virtual business	A virtual business employs electronic means to transact business as opposed to a traditional brick and mortar business that relies on face-to-face transactions with physical documents and physical currency or credit. Amazon.com was a virtual business pioneer. As an online bookstore, it delivered and brokered bookstore services without a physical retail store presence; efficiently connecting buyers and sellers without the overhead of a brick-and-mortar location.
Brainstorming	Brainstorming is a group or individual creativity technique by which efforts are made to find a conclusion for a specific problem by gathering a list of ideas spontaneously contributed by its member(s). The term was popularized by Alex Faickney Osborn in the 1963 book Applied Imagination. Osborn claimed that brainstorming was more effective than individuals working alone in generating ideas, although more recent research has questioned this conclusion.
Virtual world	A virtual world is an online community that takes the form of a computer-based simulated environment through which users can interact with one another and use and create objects. The term has become largely synonymous with interactive 3D virtual environments, where the users take the form of avatars visible to others. These avatars usually appear as textual, two-dimensional, or three-dimensional representations, although other forms are possible (auditory and touch sensations for example).
Active listening	Active listening is a communication technique that requires the listener to feed back what they hear to the speaker, by way of re-stating or paraphrasing what he has heard in his own words, to confirm what he has heard and moreover, to confirm the understanding of both parties. The ability to listen actively demonstrates sincerity, and that nothing is being assumed or taken for granted. Active listening is most often used to improve personal relationships, reduce misunderstanding and conflicts, strengthen cooperation, and foster understanding.
Receiver	In modulated ultrasound terminology, a receiver is a device that receives a modulated ultrasound signal and decodes it for use as sound, navigational-position information, etc. Its function is somewhat like that of a radio receiver

Chapter 2. Mastering Team Skills and Interpersonal Communication

Long-term memory	Long-term memory is memory in which associations among items are stored, as part of the theory of a dual-store memory model. According to the theory, long-term memory differs structurally and functionally from working memory or short-term memory, which ostensibly stores items for only around 20-30 seconds and can be recalled easily. This differs from the theory of the single-store retrieved context model that has no differentiation between short-term and long-term memory.
Online shopping	Online shopping is a form of electronic commerce whereby consumers directly buy goods or services from a seller over the Internet without an intermediary service. An online shop, eshop, e-store, Internet shop, webshop, webstore, online store, or virtual store evokes the physical analogy of buying products or services at a bricks-and-mortar retailer or shopping centre. The process is called business-to-consumer (B2C) online shopping.
Short-term memory	Short-term memory is the capacity for holding a small amount of information in mind in an active, readily available state for a short period of time. The duration of short-term memory is believed to be in the order of seconds. A commonly cited capacity is 7 ± 2 elements.
Nonverbal communication	Nonverbal communication is usually understood as the process of communication through sending and receiving wordless (mostly visual) between people. Messages can be communicated through gestures and touch, by body language or posture, by facial expression and eye contact. Nonverbal messages could also be communicated through material exponential; meaning, objects or artifacts (such as clothing, hairstyles or architecture).
Statement	In logic a statement is either (a) a meaningful declarative sentence that is either true or false, or (b) what is asserted or made by the use of a declarative sentence. In the latter case, a statement is distinct from a sentence in that a sentence is only one formulation of a statement, whereas there may be many other formulations expressing the same statement. Philosopher of language, Peter Strawson advocated the use of the term 'statement' in sense (b) in preference to proposition.
Body language	Body language is a form of mental and physical ability of human non-verbal communication, which consists of body posture, gestures, facial expressions, and eye movements. Humans send and interpret such signals almost entirely subconsciously. James Borg states that human communication consists of 93 percent body language and paralinguistic cues, while only 7% of communication consists of words themselves; however, Albert Mehrabian, the researcher whose 1960s work is the source of these statistics, has stated that this is a misunderstanding of the findings .

Chapter 2. Mastering Team Skills and Interpersonal Communication

Nonverbal	Nonverbal communication is usually understood as the process of communication through sending and receiving wordless (mostly visual) cues between people. Messages can be communicated through gestures and touch, by body language or posture, by facial expression and eye contact, which are all considered types of nonverbal communication. Speech contains nonverbal elements known as paralanguage, including voice quality, rate, pitch, volume, and speaking style, as well prosodic features such as rhythm, intonation, and stress.
Personal space	Personal space is the region surrounding a person which they regard as psychologically theirs. Most people value their personal space and feel discomfort, anger, or anxiety when their personal space is encroached. Permitting a person to enter personal space and entering somebody else's personal space are indicators of perception of the relationship between the people.
Topics	The Topics is the name given to one of Aristotle's six works on logic collectively known as the Organon. The other five are:•Categories•De Interpretatione•Prior Analytics•Posterior Analytics•On Sophistical Refutations The Topics constitutes Aristotle's treatise on the art of dialectic--the invention and discovery of arguments in which the propositions rest upon commonly-held opinions or endoxa . Topoi are 'places' from which such arguments can be discovered or invented.
Workforce management	Workforce management encompasses all the activities needed to maintain a productive workforce. Under the umbrella of human resource management, WFM is sometimes referred to as HRMS systems, or even part of ERP systems. Recently, the concept of workforce management has begun to evolve into workforce optimisation.
Appearance	In law, appearance is the coming into court of either of the parties to a lawsuit, and/or the formal act by which a defendant submits himself to the jurisdiction of the court. Legal details (outdated) The defendant in an action in the High Court of England enters his appearance to the writ of summons by delivering, either at the central office of the Supreme Court, or a district registry, a written memorandum either giving his solicitor's name or stating that he defends in person. He must also give notice to the plaintiff of his appearance, which ought, according to the time limited by the writ, to be within eight days after service; a defendant may, however, appear any time before judgment.
Business ethics	Business ethics is a form of applied ethics or professional ethics that examines ethical principles and moral or ethical problems that arise in a business environment.

	It applies to all aspects of business conduct and is relevant to the conduct of individuals and entire organizations.
	Business ethics has both normative and descriptive dimensions.
Eye contact	Eye contact is a meeting of the eyes between two individuals. In human beings, eye contact is a form of nonverbal communication and is thought to have a large influence on social behavior. Coined in the early to mid-1960s, the term has come in the West to often define the act as a meaningful and important sign of confidence and social communication.
Small business	What constitutes a small business varies widely around the world. Small businesses are normally privately owned corporations, partnerships, or sole proprietorships. What constitutes 'small' in terms of government support and tax policy varies by country and by industry, ranging from fewer than 15 employees under the Australian Fair Work Act 2009, 50 employees according to the definition used by the European Union, and fewer than 500 employees to qualify for many U.S. Small Business Administration programs, although in 2006 there were over 18,000 'small businesses' with over 500 employees that accounted for half of all the employees employed by all 'small business '.
Crowdsourcing	Crowdsourcing is a process that involves outsourcing tasks to a distributed group of people. This process can occur both online and offline. The difference between crowdsourcing and ordinary outsourcing is that a task or problem is outsourced to an undefined public rather than a specific body, such as paid employees.
Marketing Intelligence	Marketing Intelligence is the information relevant to a company's markets, gathered and analyzed specifically for the purpose of accurate and confident decision-making in determining market opportunity, market penetration strategy, and market development metrics. Marketing intelligence is necessary when entering a foreign market. Marketing Intelligence is not the same as Market Intelligence (MARKINT).

Chapter 2. Mastering Team Skills and Interpersonal Communication

1. A _____ is a business entity formed by one or more lawyers to engage in the practice of law. The primary service rendered by a _____ is to advise clients (individuals or corporations) about their legal rights and responsibilities, and to represent clients in civil or criminal cases, business transactions, and other matters in which legal advice and other assistance are sought.

    _____s are organized in a variety of ways, depending on the jurisdiction in which the firm practices.

    a. Law firm
    b. Limited company
    c. Limited liability limited partnership
    d. Limited liability partnership

2. A _____ is the formal act of giving up or quitting one's office or position. It can also refer to the act of admitting defeat in a game like chess, indicated by the resigning player declaring 'I resign', turning his king on its side, extending his hand, or stopping the chess clock. A _____ can occur when a person holding a position gained by election or appointment steps down, but leaving a position upon the expiration of a term is not considered _____.

    a. Resignation
    b. Spin geometry
    c. Differential invariant
    d. Directional derivative

3. _____ is the process of speaking to a group of people in a structured, deliberate manner intended to inform, influence, or entertain the listeners. It is closely allied to 'presenting', although the latter has more of a commercial advertisement.

    In _____, as in any form of communication, there are five basic elements, often expressed as 'who is saying what to whom using what medium with what effects?' The purpose of _____ can range from simply transmitting information, to motivating people to act, to simply telling a story.

    a. Puppet ruler
    b. Public speaking
    c. Mark Riley
    d. Term of office

4. . In logic a _____ is either (a) a meaningful declarative sentence that is either true or false, or (b) what is asserted or made by the use of a declarative sentence. In the latter case, a _____ is distinct from a sentence in that a sentence is only one formulation of a _____, whereas there may be many other formulations expressing the same _____.

    Philosopher of language, Peter Strawson advocated the use of the term '_____' in sense (b) in preference to proposition.

    a. Tacit assumption

b. Term logic

c. Testability

d. Statement

5. _____ is a term used in various branches of engineering and economic development.

_____ refers to small premises provided, often by local authorities or economic development agencies, to help new businesses to establish themselves. These typically provide not only physical space and utilities, but also administrative services and links to support and finance organisations, as well as peer support among the tenants.

a. Computer programming

b. Jimmy Carter

c. Workspace

d. Backbase

1. a
2. a
3. b
4. d
5. c

You can take the complete Chapter Practice Test

for Chapter 2. Mastering Team Skills and Interpersonal Communication
on all key terms, persons, places, and concepts.

Online 99 Cents

http://www.epub2.4.21191.2.cram101.com/

Use www.Cram101.com for all your study needs

including Cram101's online interactive problem solving labs in

chemistry, statistics, mathematics, and more.

CHAPTER OUTLINE: KEY TERMS, PEOPLE, PLACES, CONCEPTS

English

Globalization

Interactivity

Intercultural communication

Job interview

Social media

Gross domestic product

Collaborative workspace

Cultural identity

Cultural pluralism

Ethnocentrism

Intercultural

Seniority

Xenophobia

Content

Crowdsourcing

Negotiation

Request for production

Nonverbal

Nonverbal communication

Baby boomer

Gantt chart

Generation X

Generation Z

Social network

Business communication

Golden Rule

Independent contractor

Public opinion

Customs

Financial Services

Chunking

Correspondence

Financial Reporting

International business

Labor force

Proposal

User-generated content

Chapter 3. Communicating in a World of Diversity

CHAPTER OUTLINE: KEY TERMS, PEOPLE, PLACES, CONCEPTS

	Information technology
	Discovery

CHAPTER HIGHLIGHTS & NOTES: KEY TERMS, PEOPLE, PLACES, CONCEPTS

English	English is a database retrieval and reporting language somewhat like SQL, but with no programming or update abilities. It was originally released by Microdata in 1973 and named so that the company's brochures could claim that developers could generate reports on their implementation of the Pick operating system using English.
Globalization	Globalization is a common term for processes of international integration arising from increasing human connectivity and interchange of worldviews, products, ideas, and other aspects of culture. In particular, advances in transportation and telecommunications infrastructure, including the rise of the Internet, represent major driving factors in globalization and precipitate further interdependence of economic and cultural activities. Though several scholars situate the origins of globalization in modernity, others map its history long before the European age of discovery and voyages to the New World.
Interactivity	In the fields of information science, communication, and industrial design, there is debate over the meaning of interactivity. In the 'contingency view' of interactivity, there are three levels:•Noninteractive, when a message is not related to previous messages;•Reactive, when a message is related only to one immediately previous message; and•Interactive, when a message is related to a number of previous messages and to the relationship between them. Human to human communication Human communication is the basic example of interactive communication which involves two different processes; human to human interactivity and human to computer interactivity. Human-Human interactivity is the communication between people.
Intercultural communication	Intercultural communication is a form of global communication. It is used to describe the wide range of communication problems that naturally appear within an organization made up of individuals from different religious, social, ethnic, and educational backgrounds.

Chapter 3. Communicating in a World of Diversity

CHAPTER HIGHLIGHTS & NOTES: KEY TERMS, PEOPLE, PLACES, CONCEPTS

Job interview	A job interview is a process in which a potential employee is evaluated by an employer for prospective employment in their company, organization, or firm. During this process, the employer hopes to determine whether or not the applicant is suitable for the role. A job interview typically precedes the hiring decision, and is used to evaluate the candidate.
Social media	Social media includes web- and mobile-based technologies which are used to turn communication into interactive dialogue among organizations, communities, and individuals. Andreas Kaplan and Michael Haenlein define social media as 'a group of Internet-based applications that build on the ideological and technological foundations of Web 2.0, and that allow the creation and exchange of user-generated content.' When the technologies are in place, social media is ubiquitously accessible, and enabled by scalable communication techniques. Social media technologies take on many different forms including magazines, Internet forums, weblogs, social blogs, microblogging, wikis, social networks, podcasts, photographs or pictures, video, rating and social bookmarking.
Gross domestic product	Gross domestic product is the market value of all officially recognized final goods and services produced within a country in a given period. Gross domestic product per capita is often considered an indicator of a country's standard of living; Gross domestic product per capita is not a measure of personal income. Under economic theory, Gross domestic product per capita exactly equals the gross domestic income (GDI) per capita.
Collaborative workspace	A collaborative workspace is an inter-connected environment in which all the participants in dispersed locations can access and interact with each other just as inside a single entity. The environment may be supported by electronic communications and groupware which enable participants to overcome space and time differentials. These are typically enabled by a shared mental model, common information, and a shared understanding by all of the participants regardless of physical location.
Cultural identity	Cultural identity is the identity of a group or culture, or of an individual as far as one is influenced by one's belonging to a group or culture. Cultural identity is similar to and has overlaps with, but is not synonymous with, identity politics. Description Various modern cultural studies and social theories have investigated cultural identity.
Cultural pluralism	Cultural pluralism is a term used when smaller groups within a larger society maintain their unique cultural identities, and their values and practices are accepted by the wider culture provided they are consistent with the laws and values of the wider society. Cultural pluralism is often confused with Multiculturalism.

Visit Cram101.com for full Practice Exams

Ethnocentrism	Ethnocentrism is judging another culture solely by the values and standards of one's own culture. The ethnocentric individual will judge other groups relative to his or her own particular ethnic group or culture, especially with concern to language, behavior, customs, and religion. These ethnic distinctions and subdivisions serve to define each ethnicity's unique cultural identity.
Intercultural	Cross-cultural communication (also frequently referred to as intercultural communication, which is also used in a different sense, though) is a field of study that looks at how people from differing cultural backgrounds communicate, in similar and different ways among themselves, and how they endeavour to communicate across cultures. Origins

During the Cold War, the United States economy was largely self-contained because the world was polarized into two separate and competing powers: the east and west. However, changes and advancements in economic relationships, political systems, and technological options began to break down old cultural barriers. |
| Seniority | Seniority is the concept of a person or group of people being older or in charge or command of another person or group, or taking precedence over them. Control is often granted to senior persons due to experience or length of service in a given position. When persons of senior rank have less experience or length of service than their subordinates, 'seniority' may apply to either concept. |
| Xenophobia | Xenophobia is defined as an intense or irrational dislike or fear of people from other countries or as an unreasonable fear or hatred of foreigners or strangers or of that which is foreign or strange. It comes from the Greek words ξ?νος (xenos), meaning 'stranger,' 'foreigner,' and φ?βος (phobos), meaning 'fear.'

Xenophobia can manifest itself in many ways involving the relations and perceptions of an ingroup towards an outgroup, including a fear of losing identity, suspicion of its activities, aggression, and desire to eliminate its presence to secure a presumed purity. Xenophobia can also be exhibited in the form of an 'uncritical exaltation of another culture' in which a culture is ascribed 'an unreal, stereotyped and exotic quality'. |
| Content | In mathematics, a content is a real function μ defined on a field of sets $\mathcal{A}$ such that $$\mu(A) \in [0, \infty] \text{ whenever } A \in \mathcal{A}. \; \mu(\varnothing) = 0.$$ $\mu(A_1 \cup A_2) = \mu(A_1) + \mu(A_2)$ whenever $A_1, A_2 \in \mathcal{A}$ and $A_1 \cap A_2 = \varnothing$.

A very important type of content is a measure, which is a σ-additive content defined on a σ-field. Every measure is a content, but not vice-versa. |
| Crowdsourcing | Crowdsourcing is a process that involves outsourcing tasks to a distributed group of people. |

	This process can occur both online and offline. The difference between crowdsourcing and ordinary outsourcing is that a task or problem is outsourced to an undefined public rather than a specific body, such as paid employees.
Negotiation	Negotiation is a dialogue between two or more people or parties, intended to reach an understanding, resolve point of difference, or gain advantage in outcome of dialogue, to produce an agreement upon courses of action, to bargain for individual or collective advantage, to craft outcomes to satisfy various interests of two people/parties involved in negotiation process. Negotiation is a process where each party involved in negotiating tries to gain an advantage for themselves by the end of the process. Negotiation is intended to aim at compromise.
Request for production	A request for production is a legal request for documents, electronically stored information, or other tangible items. In civil procedure, during the discovery phase of litigation, a party to a lawsuit may request that another party provide any documents that it has that pertain to the subject matter of the lawsuit. For example, a party in a court case may obtain copies of e-mail messages sent by employees of the opposing party.
Nonverbal	Nonverbal communication is usually understood as the process of communication through sending and receiving wordless (mostly visual) cues between people. Messages can be communicated through gestures and touch, by body language or posture, by facial expression and eye contact, which are all considered types of nonverbal communication. Speech contains nonverbal elements known as paralanguage, including voice quality, rate, pitch, volume, and speaking style, as well prosodic features such as rhythm, intonation, and stress.
Nonverbal communication	Nonverbal communication is usually understood as the process of communication through sending and receiving wordless (mostly visual) between people. Messages can be communicated through gestures and touch, by body language or posture, by facial expression and eye contact. Nonverbal messages could also be communicated through material exponential; meaning, objects or artifacts (such as clothing, hairstyles or architecture).
Baby boomer	A baby boomer is a person who was born during the demographic Post-World War II baby boom between the years 1946 and 1964, according to the U.S. Census Bureau. The term 'baby boomer' is sometimes used in a cultural context. Therefore, it is impossible to achieve broad consensus of a precise definition, even within a given territory.
Gantt chart	A Gantt chart is a type of bar chart, developed by Henry Gantt, that illustrates a project schedule. Gantt charts illustrate the start and finish dates of the terminal elements and summary elements of a project.

Generation X	Generation X, commonly abbreviated to Gen X, is the generation born after the Western post-World War II baby boom ended. While there is no universally agreed upon time frame, commentators usually use beginning birth dates ranging from the early 1960s to the early 1980s . The term came to be popularized by Douglas Coupland's 1991 book of the same name, and had also been used in different times and places for various subcultures or countercultures since the 1950s.
Generation Z	Generation Z is a common name for the group of people born from a currently undefined point in the last decade of the 20th century and the beginning of the 21st century, from the early years of the 1990s decade through to the present, as distinct from the preceding 'Generation Y' (also referred to as 'Millennials').
	The most recent cultural generation refers to those born beginning in either the early-1990s or early 2000s, some sources including those born as early as 1989-1991. Most members of the generation have spent their entire lives with the World Wide Web, having been commercially available since 1991, which has proved to be a major influence. The youngest of the generation were born during a baby boomlet around the time of the Global financial crisis of the late 2000s, ending around the year 2010, with the next unnamed generation succeeding.
Social network	A social network is a social structure made up of a set of actors (such as individuals or organizations) and the dyadic ties between these actors. The social network perspective provides a clear way of analyzing the structure of whole social entities. The study of these structures uses social network analysis to identify local and global patterns, locate influential entities, and examine network dynamics.
Business communication	Business Communication: communication used to promote a product, service, or organization; relay information within the business; or deal with legal and similar issues. It is also a means of relaying between a supply chain, for example the consumer and manufacturer.
	Business Communication is known simply as 'communications'.
Golden Rule	The Golden Rule is a maxim, ethical code, or morality that essentially states either of the following:•(Positive form): One should treat others as one would like others to treat oneself.•(Negative/prohibitive form, also called the Silver Rule): One should not treat others in ways that one would not like to be treated.
	This concept describes a 'reciprocal' or 'two-way' relationship between one's self and others that involves both sides equally and in a mutual fashion.
	This concept can be explained from the perspective of psychology, philosophy, sociology, religion, etc.: Psychologically it involves a person empathizing with others.

Chapter 3. Communicating in a World of Diversity

Independent contractor	An independent contractor is a natural person, business, or corporation that provides goods or services to another entity under terms specified in a contract or within a verbal agreement. Unlike an employee, an independent contractor does not work regularly for an employer but works as and when required, during which time he or she may be subject to the Law of Agency. Independent contractors are usually paid on a freelance basis.
Public opinion	Public opinion is the aggregate of individual attitudes or beliefs held by the adult population. Public opinion can also be defined as the complex collection of opinions of many different people and the sum of all their views.
	The principle approaches to the study of public opinion may be divided into 4 categories:•quantitative measurement of opinion distributions;•investigation of the internal relationships among the individual opinions that make up public opinion on an issue;•description or analysis of the public role of public opinion;•study both of the communication media that disseminate the ideas on which opinions are based and of the uses that propagandists and other manipulators make of these media.Concepts of public opinion
	Public opinion as a concept gained credence with the rise of 'public' in the eighteenth century.
Customs	Customs is an authority or agency in a country responsible for collecting and safeguarding customs duties and for controlling the flow of goods including animals, transports, personal effects and hazardous items in and out of a country. Depending on local legislation and regulations, the import or export of some goods may be restricted or forbidden, and the customs agency enforces these rules. The customs authority may be different from the immigration authority, which monitors persons who leave or enter the country, checking for appropriate documentation, apprehending people wanted by international arrest warrants, and impeding the entry of others deemed dangerous to the country.
Financial Services	Financial services are the economic services provided by the finance industry, which encompasses a broad range of organizations that manage money, including credit unions, banks, credit card companies, insurance companies, consumer finance companies, stock brokerages, investment funds and some government sponsored enterprises. As of 2004, the financial services industry represented 20% of the market capitalization of the S&P 500 in the United States. History of financial services
	The term 'financial services' became more prevalent in the United States partly as a result of the Gramm-Leach-Bliley Act of the late 1990s, which enabled different types of companies operating in the U.S. financial services industry at that time to merge.

Chunking	In mathematics education at primary school level, chunking (sometimes also called the partial quotients method) is an elementary approach for solving simple division questions, by repeated subtraction.
	To calculate the result of dividing a large number by a small number, the student repeatedly takes away 'chunks' of the large number, where each 'chunk' is an easy multiple (for example 100×, 10×, 5× 2×, etc). of the small number, until the large number has been reduced to zero or the remainder is less than the divisor.
Correspondence	In mathematics and mathematical economics, correspondence is a term with several related but not identical meanings. •In general mathematics, a correspondence is an ordered triple (X,Y,R), where R is a relation from X to Y, i.e. any subset of the Cartesian product X×Y.•In algebraic geometry, a correspondence between algebraic varieties V and W is in the same fashion a subset R of V×W, which is in addition required to be closed in the Zariski topology. It therefore means any relation that is defined by algebraic equations.
Financial Reporting	Financial reporting is the process of preparing and distributing financial information to users of such information in various forms. The most common format of formal financial reporting are financial statements. Financial statements are prepared in accordance with rigorously applied standards defined by professional accounting bodies developed according to the legal and professional framework of a specific locale.
International business	International business is a term used to collectively describe all commercial transactions (private and governmental, sales, investments, logistics,and transportation) that take place between two or more regions, countries and nations beyond their political boundary. Usually, private companies undertake such transactions for profit; governments undertake them for profit and for political reasons. It refers to all those business activities which involve cross border transactions of goods, services, resources between two or more nations.
Labor force	Normally, the labor force of a country consists of everyone of working age (typically above a certain age (around 14 to 16) and below retirement (around 65) who are participating workers, that is people actively employed or seeking employment. People not counted include students, retired people, stay-at-home parents, people in prisons or similar institutions, people employed in jobs or professions with unreported income, as well as discouraged workers who cannot find work.
	In the United States, the unemployment rate is estimated by a household survey called the Current Population Survey, conducted monthly by the Federal Bureau of Labor Statistics.
Proposal	A business proposal is a written offer from a seller to a prospective buyer.

Chapter 3. Communicating in a World of Diversity

Business proposals are often a key step in the complex sales process--i.e., whenever a buyer considers more than price in a purchase.

There are three distinct categories of business proposals:•formally solicited•informally solicited•unsolicited.

Solicited proposals are written in response to published requirements, contained in a Request for Proposal (RFP), Request for Quotation (RFQ), Request for Information (RFI) or an Invitation For Bid (IFB).

| User-generated content | User-generated content covers a range of media content available in a range of modern communications technologies. It entered mainstream usage during 2005, having arisen in web publishing and new media content production circles. Its use for a wide range of applications, including problem processing, news, gossip and research, reflects the expansion of media production through new technologies that are accessible and affordable to the general public. |

| Information technology | Information technology is a branch of engineering dealing with the use of computers and telecommunications equipment to store, retrieve, transmit and manipulate data. The term is commonly used as a synonym for computers and computer networks, but it also encompasses other information distribution technologies such as television and telephones. Some of the modern and emerging fields of Information technology are next generation web technologies, bioinformatics, cloud computing, global information systems, large-scale knowledge bases, etc. |

| Discovery | Discovery is the act of detecting something new, or something 'old' that had been unknown. With reference to science and academic disciplines, discovery is the observation of new phenomena, new actions, or new events and providing new reasoning to explain the knowledge gathered through such observations with previously acquired knowledge from abstract thought and everyday experiences. Visual discoveries are often called sightings. |

1. _____ is defined as an intense or irrational dislike or fear of people from other countries or as an unreasonable fear or hatred of foreigners or strangers or of that which is foreign or strange. It comes from the Greek words ξ?νος (xenos), meaning 'stranger,' 'foreigner,' and φ?βος (phobos), meaning 'fear.'

   _____ can manifest itself in many ways involving the relations and perceptions of an ingroup towards an outgroup, including a fear of losing identity, suspicion of its activities, aggression, and desire to eliminate its presence to secure a presumed purity. _____ can also be exhibited in the form of an 'uncritical exaltation of another culture' in which a culture is ascribed 'an unreal, stereotyped and exotic quality'.

   a. Comet
   b. Xenophobia
   c. symptom
   d. Climate

2. In the fields of information science, communication, and industrial design, there is debate over the meaning of _____. In the 'contingency view' of _____, there are three levels:•Noninteractive, when a message is not related to previous messages;•Reactive, when a message is related only to one immediately previous message; and•Interactive, when a message is related to a number of previous messages and to the relationship between them. Human to human communication

   Human communication is the basic example of interactive communication which involves two different processes; human to human _____ and human to computer _____. Human-Human _____ is the communication between people.

   a. Edith of Wilton
   b. Retraining
   c. Kennedy Round
   d. Interactivity

3. A _____ is a person who was born during the demographic Post-World War II baby boom between the years 1946 and 1964, according to the U.S. Census Bureau. The term '_____' is sometimes used in a cultural context. Therefore, it is impossible to achieve broad consensus of a precise definition, even within a given territory.

   a. Jimmy Carter
   b. Multifactor
   c. Simple interest
   d. Baby boomer

4. . _____: communication used to promote a product, service, or organization; relay information within the business; or deal with legal and similar issues. It is also a means of relaying between a supply chain, for example the consumer and manufacturer.

   _____ is known simply as 'communications'.

    a. Business communication

    b. Business Motivation Model

    c. Business Process Definition Metamodel

    d. Business process management

5. A _____ is a process in which a potential employee is evaluated by an employer for prospective employment in their company, organization, or firm. During this process, the employer hopes to determine whether or not the applicant is suitable for the role.

    A _____ typically precedes the hiring decision, and is used to evaluate the candidate.

    a. Jimmy Carter

    b. Retraining

    c. Kennedy Round

    d. Job interview

1. b
2. d
3. d
4. a
5. d

CHAPTER OUTLINE: KEY TERMS, PEOPLE, PLACES, CONCEPTS

Job interview

Writing style

Survey research

Audience analysis

Public opinion

Statement

Discovery

Information needs

Cultural identity

Form letter

Proposal

Search Engine

Social networking

Virtual team

Electronic business

Information graphics

YouTube

Formality

Interactive media

	Cost
	Brainstorming
	Intercultural
	Intercultural communication
	Storyboard
	Topics
	Content
	Independent contractor
	User-generated content

CHAPTER HIGHLIGHTS & NOTES: KEY TERMS, PEOPLE, PLACES, CONCEPTS

Job interview	A job interview is a process in which a potential employee is evaluated by an employer for prospective employment in their company, organization, or firm. During this process, the employer hopes to determine whether or not the applicant is suitable for the role.
	A job interview typically precedes the hiring decision, and is used to evaluate the candidate.
Writing style	Writing style is the manner in which an author chooses to write to his or her audience. A style reveals both the writer's personality and voice, but it also shows how she or he perceives the audience, and chooses conceptual writing style which reveal those choices by which the writer may change the conceptual world of the overall character of the work. This might be done by a simple change of words; a syntactical structure, parsing prose, adding diction, and organizing figures of thought into usable frameworks.

Survey research	Survey research involves utilizing interviews or questionnaires to obtain quantitative information in fields such as marketing, politics, and social science. Utilizing surveys is considered to be an efficient way of collecting data from a large number of respondents, accurately representing a whole population. Surveys also have the benefit of providing data that is relatively free from errors.
Audience analysis	Audience analysis is a task that is often performed by technical writers in a project's early stages. It consists of assessing the audience to make sure the information provided to them is at the appropriate level. The audience is often referred to as the end-user, and all communications need to be targeted towards the defined audience.
Public opinion	Public opinion is the aggregate of individual attitudes or beliefs held by the adult population. Public opinion can also be defined as the complex collection of opinions of many different people and the sum of all their views. The principle approaches to the study of public opinion may be divided into 4 categories:•quantitative measurement of opinion distributions;•investigation of the internal relationships among the individual opinions that make up public opinion on an issue;•description or analysis of the public role of public opinion;•study both of the communication media that disseminate the ideas on which opinions are based and of the uses that propagandists and other manipulators make of these media.Concepts of public opinion Public opinion as a concept gained credence with the rise of 'public' in the eighteenth century.
Statement	In logic a statement is either (a) a meaningful declarative sentence that is either true or false, or (b) what is asserted or made by the use of a declarative sentence. In the latter case, a statement is distinct from a sentence in that a sentence is only one formulation of a statement, whereas there may be many other formulations expressing the same statement. Philosopher of language, Peter Strawson advocated the use of the term 'statement' in sense (b) in preference to proposition.
Discovery	Discovery is the act of detecting something new, or something 'old' that had been unknown. With reference to science and academic disciplines, discovery is the observation of new phenomena, new actions, or new events and providing new reasoning to explain the knowledge gathered through such observations with previously acquired knowledge from abstract thought and everyday experiences. Visual discoveries are often called sightings.
Information needs	Information need is an individual or group's desire to locate and obtain information to satisfy a conscious or unconscious need. The 'information' and 'need' in 'information need' are inseparable interconnection.

Chapter 4. Planning Business Messages

Cultural identity	Cultural identity is the identity of a group or culture, or of an individual as far as one is influenced by one's belonging to a group or culture. Cultural identity is similar to and has overlaps with, but is not synonymous with, identity politics. Description Various modern cultural studies and social theories have investigated cultural identity.
Form letter	A form letter is a letter written from a template, rather than being specially composed for a specific recipient. The most general kind of form letter consists of one or more regions of boilerplate text interspersed with one or more substitution placeholders. Although form letters are generally intended for a wide audience, many form letters include stylistic elements or features intended to appear specifically tailored to the recipient -- for example, they might be signed by autopen and utilize features such as mail merge to automatically insert the names of individual recipients.
Proposal	A business proposal is a written offer from a seller to a prospective buyer. Business proposals are often a key step in the complex sales process--i.e., whenever a buyer considers more than price in a purchase. There are three distinct categories of business proposals:•formally solicited•informally solicited•unsolicited. Solicited proposals are written in response to published requirements, contained in a Request for Proposal (RFP), Request for Quotation (RFQ), Request for Information (RFI) or an Invitation For Bid (IFB).
Search Engine	Search Engine was a weekly Canadian radio show that aired on CBC Radio One, then a dedicated podcast available on CBC.ca, and now a podcast on TVOntario's website, tvo.org. It is hosted by Jesse Brown, who also co-produces the show with Geoff Siskind and Andrew Parker. Cory Doctorow, novelist and editor of Boing Boing, is also a regular contributor.
Social networking	A social networking service is an online service, platform, or site that focuses on facilitating the building of social networks or social relations among people who, for example, share interests, activities, backgrounds, or real-life connections. A social network service consists of a representation of each user (often a profile), his/her social links, and a variety of additional services. Most social network services are web-based and provide means for users to interact over the Internet, such as e-mail and instant messaging.
Virtual team	A virtual team is a group of individuals who work across time, space and organizational boundaries with links strengthened by webs of communication technology' Ale Ebrahim, N., Ahmed, S. & Taha, Z.

	in their recent (2009) literature review paper, added two key issues to definition of a virtual team 'as small temporary groups of geographically, organizationally and/ or time dispersed knowledge workers who coordinate their work predominantly with electronic information and communication technologies in order to accomplish one or more organization tasks' Members of virtual teams communicate electronically and may never meet face-to-face. Virtual teams are made possible by a proliferation of fiber optic technology that has significantly increased the scope of off-site communication.
Electronic business	Electronic business, commonly referred to as 'eBusiness' or 'e-business', or an internet business, may be defined as the application of information and communication technologies (ICT) in support of all the activities of business. Commerce constitutes the exchange of products and services between businesses, groups and individuals and can be seen as one of the essential activities of any business. Electronic commerce focuses on the use of ICT to enable the external activities and relationships of the business with individuals, groups and other businesses.
Information graphics	Information graphics are graphic visual representations of information, data or knowledge. These graphics present complex information quickly and clearly, such as in signs, maps, journalism, technical writing, and education. With an information graphic, computer scientists, mathematicians, and statisticians develop and communicate concepts using a single symbol to process information.
YouTube	YouTube is a video-sharing website, created by three former PayPal employees in February 2005, on which users can upload, view and share videos. The company is based in San Bruno, California, and uses Adobe Flash Video and HTML5 technology to display a wide variety of user-generated video content, including movie clips, TV clips, and music videos, as well as amateur content such as video blogging, short original videos, and educational videos. Most of the content on YouTube has been uploaded by individuals, although media corporations including CBS, the BBC, VEVO, Hulu, and other organizations offer some of their material via the site, as part of the YouTube partnership program.
Formality	Utterances, conceptually similar to a ritual although typically secular and less involved. A formality may be as simple as a handshake upon making new acquaintinces in Western culture to the carefully defined procedure of bows, handshakes, formal greetings, and business-card exchanges that may mark two businessmen being introduced in Japan. In legal and diplomatic circles, formalities include such matters as greeting an arriving head of state with the appropriate national anthem.

Chapter 4. Planning Business Messages

Interactive media	Interactive media normally refers to products and services on digital computer-based systems which respond to the user's actions by presenting content such as text, graphics, animation, video, audio, games, etc. Though the word media is plural, the term is often used as a singular noun. Interactive media is related to the concepts interaction design, new media, interactivity, human computer interaction, cyberculture, digital culture, and includes specific cases such as, for example, interactive television, interactive narrative, interactive advertising, algorithmic art, videogames, social media, ambient intelligence, virtual reality and augmented reality.
Cost	In production, research, retail, and accounting, a cost is the value of money that has been used up to produce something, and hence is not available for use anymore. In business, the cost may be one of acquisition, in which case the amount of money expended to acquire it is counted as cost. In this case, money is the input that is gone in order to acquire the thing.
Brainstorming	Brainstorming is a group or individual creativity technique by which efforts are made to find a conclusion for a specific problem by gathering a list of ideas spontaneously contributed by its member(s). The term was popularized by Alex Faickney Osborn in the 1963 book Applied Imagination. Osborn claimed that brainstorming was more effective than individuals working alone in generating ideas, although more recent research has questioned this conclusion.
Intercultural	Cross-cultural communication (also frequently referred to as intercultural communication, which is also used in a different sense, though) is a field of study that looks at how people from differing cultural backgrounds communicate, in similar and different ways among themselves, and how they endeavour to communicate across cultures. Origins During the Cold War, the United States economy was largely self-contained because the world was polarized into two separate and competing powers: the east and west. However, changes and advancements in economic relationships, political systems, and technological options began to break down old cultural barriers.
Intercultural communication	Intercultural communication is a form of global communication. It is used to describe the wide range of communication problems that naturally appear within an organization made up of individuals from different religious, social, ethnic, and educational backgrounds. Intercultural communication is sometimes used synonymously with cross-cultural communication.
Storyboard	Storyboards are graphic organizers in the form of illustrations or images displayed in sequence for the purpose of pre-visualizing a motion picture, animation, motion graphic or interactive media sequence.

	The storyboarding process, in the form it is known today, was developed at the Walt Disney Studio during the early 1930s, after several years of similar processes being in use at Walt Disney and other animation studios. Origins
	The storyboarding process can be very time-consuming and intricate.
Topics	The Topics is the name given to one of Aristotle's six works on logic collectively known as the Organon. The other five are:•Categories•De Interpretatione•Prior Analytics•Posterior Analytics•On Sophistical Refutations
	The Topics constitutes Aristotle's treatise on the art of dialectic--the invention and discovery of arguments in which the propositions rest upon commonly-held opinions or endoxa . Topoi are 'places' from which such arguments can be discovered or invented.
Content	In mathematics, a content is a real function μ defined on a field of sets $\mathcal{A}$ such that• $$\mu(A) \in [0, \infty] \text{ whenever } A \in \mathcal{A}. \, \mu(\varnothing) = 0. \,$$ $\mu(A_1 \cup A_2) = \mu(A_1) + \mu(A_2)$ whenever $A_1, A_2 \in \mathcal{A}$ and $A_1 \cap A_2 = \varnothing$.
	A very important type of content is a measure, which is a σ-additive content defined on a σ-field. Every measure is a content, but not vice-versa.
Independent contractor	An independent contractor is a natural person, business, or corporation that provides goods or services to another entity under terms specified in a contract or within a verbal agreement. Unlike an employee, an independent contractor does not work regularly for an employer but works as and when required, during which time he or she may be subject to the Law of Agency. Independent contractors are usually paid on a freelance basis.
User-generated content	User-generated content covers a range of media content available in a range of modern communications technologies. It entered mainstream usage during 2005, having arisen in web publishing and new media content production circles. Its use for a wide range of applications, including problem processing, news, gossip and research, reflects the expansion of media production through new technologies that are accessible and affordable to the general public.

Chapter 4. Planning Business Messages

1. _____ normally refers to products and services on digital computer-based systems which respond to the user's actions by presenting content such as text, graphics, animation, video, audio, games, etc.

   Though the word media is plural, the term is often used as a singular noun.

   _____ is related to the concepts interaction design, new media, interactivity, human computer interaction, cyberculture, digital culture, and includes specific cases such as, for example, interactive television, interactive narrative, interactive advertising, algorithmic art, videogames, social media, ambient intelligence, virtual reality and augmented reality.

   a. Edith of Wilton
   b. Interactive media
   c. EPAS
   d. Interface Technologies

2. Utterances, conceptually similar to a ritual although typically secular and less involved. A _____ may be as simple as a handshake upon making new acquaintances in Western culture to the carefully defined procedure of bows, handshakes, formal greetings, and business-card exchanges that may mark two businessmen being introduced in Japan. In legal and diplomatic circles, _____(ies) include such matters as greeting an arriving head of state with the appropriate national anthem.

   a. Jimmy Carter
   b. EMortgages
   c. Formality
   d. Interface Technologies

3. A _____ is a group of individuals who work across time, space and organizational boundaries with links strengthened by webs of communication technology' Ale Ebrahim, N., Ahmed, S. & Taha, Z. in their recent (2009) literature review paper, added two key issues to definition of a _____ 'as small temporary groups of geographically, organizationally and/ or time dispersed knowledge workers who coordinate their work predominantly with electronic information and communication technologies in order to accomplish one or more organization tasks' Members of _____s communicate electronically and may never meet face-to-face. _____s are made possible by a proliferation of fiber optic technology that has significantly increased the scope of off-site communication.

   a. Personalized marketing
   b. Jimmy Carter
   c. Virtual team
   d. Tips from the Top Floor

4. . _____(s) is an individual or group's desire to locate and obtain information to satisfy a conscious or unconscious need. The 'information' and 'need' in '_____(s)' are inseparable interconnection. Needs and interests call forth information.

a. Information needs

b. Dreyfus model of skill acquisition

c. Dysrationalia

d. Habituation

5. _____ is the manner in which an author chooses to write to his or her audience. A style reveals both the writer's personality and voice, but it also shows how she or he perceives the audience, and chooses conceptual _____ which reveal those choices by which the writer may change the conceptual world of the overall character of the work. This might be done by a simple change of words; a syntactical structure, parsing prose, adding diction, and organizing figures of thought into usable frameworks.

a. Writer's block

b. Writing style

c. verbosity

d. Jimmy Carter

1. b
2. c
3. c
4. a
5. b

You can take the complete Chapter Practice Test

for Chapter 4. Planning Business Messages
on all key terms, persons, places, and concepts.

Online 99 Cents

http://www.epub2.4.21191.4.cram101.com/

Use www.Cram101.com for all your study needs

including Cram101's online interactive problem solving labs in

chemistry, statistics, mathematics, and more.

Chapter 5. Writing Business Messages

Intellectual property

Audience analysis

Cultural identity

Social networking

Job interview

Labeling

Labor force

Gantt chart

Credential

Credibility

Objectivity

Social media

Intercultural

Intercultural communication

Storyboard

Formality

English

Content

Interactivity

Chapter 5. Writing Business Messages

_____ | P3M3

_____ | Request for production

_____ | Collaborative workspace

_____ | Information technology

_____ | Dependent clause

_____ | Topics

_____ | Independent contractor

_____ | Mail merge

_____ | Public opinion

_____ | Pull quote

_____ | Search Engine

CHAPTER HIGHLIGHTS & NOTES: KEY TERMS, PEOPLE, PLACES, CONCEPTS

Intellectual property	Intellectual property is a term referring to a number of distinct types of expressions for which a set of rights are recognized under the corresponding fields of law. Under intellectual property law, owners are granted certain exclusive rights to various markets, machines, musical, literary, and artistic works; discoveries and inventions; and applications. Common types of intellectual property rights include copyrights, trademarks, patents, industrial design rights, and trade secrets in some jurisdictions.
Audience analysis	Audience analysis is a task that is often performed by technical writers in a project's early stages. It consists of assessing the audience to make sure the information provided to them is at the appropriate level.

Chapter 5. Writing Business Messages

Cultural identity	Cultural identity is the identity of a group or culture, or of an individual as far as one is influenced by one's belonging to a group or culture. Cultural identity is similar to and has overlaps with, but is not synonymous with, identity politics. Description

Various modern cultural studies and social theories have investigated cultural identity. |
| Social networking | A social networking service is an online service, platform, or site that focuses on facilitating the building of social networks or social relations among people who, for example, share interests, activities, backgrounds, or real-life connections. A social network service consists of a representation of each user (often a profile), his/her social links, and a variety of additional services. Most social network services are web-based and provide means for users to interact over the Internet, such as e-mail and instant messaging. |
| Job interview | A job interview is a process in which a potential employee is evaluated by an employer for prospective employment in their company, organization, or firm. During this process, the employer hopes to determine whether or not the applicant is suitable for the role.

A job interview typically precedes the hiring decision, and is used to evaluate the candidate. |
| Labeling | Cartographic labeling is a form of typography and strongly deals with form, style, weight and size of type on a map. Essentially, labeling denotes the correct way to label features (points, arcs, or polygons).

In type, form describes anything from lengths between letters to the case and color of the font. |
| Labor force | Normally, the labor force of a country consists of everyone of working age (typically above a certain age (around 14 to 16) and below retirement (around 65) who are participating workers, that is people actively employed or seeking employment. People not counted include students, retired people, stay-at-home parents, people in prisons or similar institutions, people employed in jobs or professions with unreported income, as well as discouraged workers who cannot find work.

In the United States, the unemployment rate is estimated by a household survey called the Current Population Survey, conducted monthly by the Federal Bureau of Labor Statistics. |
| Gantt chart | A Gantt chart is a type of bar chart, developed by Henry Gantt, that illustrates a project schedule. Gantt charts illustrate the start and finish dates of the terminal elements and summary elements of a project. Terminal elements and summary elements comprise the work breakdown structure of the project. |

Credential	A credential is an attestation of qualification, competence, or authority issued to an individual by a third party with a relevant or de facto authority or assumed competence to do so.
	Examples of credentials include academic diplomas, academic degrees, certifications, security clearances, identification documents, badges, passwords, user names, keys, powers of attorney, and so on. Sometimes publications, such as scientific papers or books, may be viewed as similar to credentials by some people, especially if the publication was peer reviewed or made in a well-known journal or reputable publisher.
Credibility	Credibility refers to the objective and subjective components of the believability of a source or message.
	Traditionally, modern, credibility has two key components: trustworthiness and expertise, which both have objective and subjective components. Trustworthiness is based more on subjective factors, but can include objective measurements such as established reliability.
Objectivity	Objectivity in science is a value that informs how science is practiced and how scientific truths are created. It is the idea that scientists, in attempting to uncover truths about the natural world, must aspire to eliminate personal biases, a priori commitments, emotional involvement, etc. Objectivity is often attributed to the property of scientific measurement, as the accuracy of a measurement can be tested independent from the individual scientist who first reports it.
Social media	Social media includes web- and mobile-based technologies which are used to turn communication into interactive dialogue among organizations, communities, and individuals. Andreas Kaplan and Michael Haenlein define social media as 'a group of Internet-based applications that build on the ideological and technological foundations of Web 2.0, and that allow the creation and exchange of user-generated content.' When the technologies are in place, social media is ubiquitously accessible, and enabled by scalable communication techniques.
	Social media technologies take on many different forms including magazines, Internet forums, weblogs, social blogs, microblogging, wikis, social networks, podcasts, photographs or pictures, video, rating and social bookmarking.
Intercultural	Cross-cultural communication (also frequently referred to as intercultural communication, which is also used in a different sense, though) is a field of study that looks at how people from differing cultural backgrounds communicate, in similar and different ways among themselves, and how they endeavour to communicate across cultures. Origins
	During the Cold War, the United States economy was largely self-contained because the world was polarized into two separate and competing powers: the east and west.

Chapter 5. Writing Business Messages

Intercultural communication	Intercultural communication is a form of global communication. It is used to describe the wide range of communication problems that naturally appear within an organization made up of individuals from different religious, social, ethnic, and educational backgrounds. Intercultural communication is sometimes used synonymously with cross-cultural communication.
Storyboard	Storyboards are graphic organizers in the form of illustrations or images displayed in sequence for the purpose of pre-visualizing a motion picture, animation, motion graphic or interactive media sequence.
	The storyboarding process, in the form it is known today, was developed at the Walt Disney Studio during the early 1930s, after several years of similar processes being in use at Walt Disney and other animation studios. Origins
	The storyboarding process can be very time-consuming and intricate.
Formality	Utterances, conceptually similar to a ritual although typically secular and less involved. A formality may be as simple as a handshake upon making new acquaintaences in Western culture to the carefully defined procedure of bows, handshakes, formal greetings, and business-card exchanges that may mark two businessmen being introduced in Japan. In legal and diplomatic circles, formalities include such matters as greeting an arriving head of state with the appropriate national anthem.
English	English is a database retrieval and reporting language somewhat like SQL, but with no programming or update abilities. It was originally released by Microdata in 1973 and named so that the company's brochures could claim that developers could generate reports on their implementation of the Pick operating system using English.
Content	In mathematics, a content is a real function μ defined on a field of sets $\mathcal{A}$ such that• $$\mu(A) \in [0, \infty] \text{ whenever } A \in \mathcal{A}. \; \mu(\varnothing) = 0. \;$$ $\mu(A_1 \cup A_2) = \mu(A_1) + \mu(A_2)$ whenever $A_1, A_2 \in \mathcal{A}$ and $A_1 \cap A_2 = \varnothing$.
	A very important type of content is a measure, which is a σ-additive content defined on a σ-field. Every measure is a content, but not vice-versa.
Interactivity	In the fields of information science, communication, and industrial design, there is debate over the meaning of interactivity. In the 'contingency view' of interactivity, there are three levels:•Noninteractive, when a message is not related to previous messages;•Reactive, when a message is related only to one immediately previous message; and•Interactive, when a message is related to a number of previous messages and to the relationship between them. Human to human communication

	Human communication is the basic example of interactive communication which involves two different processes; human to human interactivity and human to computer interactivity. Human-Human interactivity is the communication between people.
P3M3	P3M3, Programme and Project Management Maturity Model is a reference guide for structured best practice. It breaks down the broad disciplines of portfolio, programme and project management into a hierarchy of Key Process Areas (KPAs). The hierarchical approach enables organisations to assess their current capability and then plot a roadmap for improvement prioritised by those KPAs which will make the biggest impact on performance.
Request for production	A request for production is a legal request for documents, electronically stored information, or other tangible items. In civil procedure, during the discovery phase of litigation, a party to a lawsuit may request that another party provide any documents that it has that pertain to the subject matter of the lawsuit. For example, a party in a court case may obtain copies of e-mail messages sent by employees of the opposing party.
Collaborative workspace	A collaborative workspace is an inter-connected environment in which all the participants in dispersed locations can access and interact with each other just as inside a single entity. The environment may be supported by electronic communications and groupware which enable participants to overcome space and time differentials. These are typically enabled by a shared mental model, common information, and a shared understanding by all of the participants regardless of physical location.
Information technology	Information technology is a branch of engineering dealing with the use of computers and telecommunications equipment to store, retrieve, transmit and manipulate data. The term is commonly used as a synonym for computers and computer networks, but it also encompasses other information distribution technologies such as television and telephones. Some of the modern and emerging fields of Information technology are next generation web technologies, bioinformatics, cloud computing, global information systems, large-scale knowledge bases, etc.
Dependent clause	In linguistics, a dependent clause is a clause that augments an independent clause with additional information, but which cannot stand alone as a sentence. Dependent clauses modify the independent clause of a sentence or serve as a component of it. Some grammarians use the term subordinate clause as a synonym for dependent clause, but in some grammars subordinate clause refers only to adverbial dependent clauses.There are also different types of dependent clauses like noun clauses, relative (adjectival) clauses, and adverbial clauses.
Topics	The Topics is the name given to one of Aristotle's six works on logic collectively known as the Organon.

	The other five are:•Categories•De Interpretatione•Prior Analytics•Posterior Analytics•On Sophistical Refutations The Topics constitutes Aristotle's treatise on the art of dialectic--the invention and discovery of arguments in which the propositions rest upon commonly-held opinions or endoxa . Topoi are 'places' from which such arguments can be discovered or invented.
Independent contractor	An independent contractor is a natural person, business, or corporation that provides goods or services to another entity under terms specified in a contract or within a verbal agreement. Unlike an employee, an independent contractor does not work regularly for an employer but works as and when required, during which time he or she may be subject to the Law of Agency. Independent contractors are usually paid on a freelance basis.
Mail merge	Mail merge is a software function describing the production of multiple (and potentially large numbers of) documents from a single template form and a structured data source. MailMerge was originally the name of a utility supplied with MultiMate an early word processing product for the IBM PC, patterned after Wang word processors. This technique of merging data to create gave rise to the term mail merge.
Public opinion	Public opinion is the aggregate of individual attitudes or beliefs held by the adult population. Public opinion can also be defined as the complex collection of opinions of many different people and the sum of all their views. The principle approaches to the study of public opinion may be divided into 4 categories:•quantitative measurement of opinion distributions;•investigation of the internal relationships among the individual opinions that make up public opinion on an issue;•description or analysis of the public role of public opinion;•study both of the communication media that disseminate the ideas on which opinions are based and of the uses that propagandists and other manipulators make of these media.Concepts of public opinion Public opinion as a concept gained credence with the rise of 'public' in the eighteenth century.
Pull quote	A pull quote is a quotation or excerpt from an article that is typically placed in a larger or distinctive typeface on the same page, serving to entice readers into an article or to highlight a key topic. The term is principally used in journalism and publishing. Placement of a pull quote on a page may be defined in a publication's or website's style guide.
Search Engine	Search Engine was a weekly Canadian radio show that aired on CBC Radio One, then a dedicated podcast available on CBC.ca, and now a podcast on TVOntario's website, tvo.org.

It is hosted by Jesse Brown, who also co-produces the show with Geoff Siskind and Andrew Parker. Cory Doctorow, novelist and editor of Boing Boing, is also a regular contributor.

1. _____ is a term referring to a number of distinct types of expressions for which a set of rights are recognized under the corresponding fields of law. Under _____ law, owners are granted certain exclusive rights to various markets, machines, musical, literary, and artistic works; discoveries and inventions; and applications. Common types of _____ rights include copyrights, trademarks, patents, industrial design rights, and trade secrets in some jurisdictions.

   a. organizational learning
   b. Information economy
   c. Internet forum
   d. Intellectual property

2. In mathematics, a _____ is a real function μ defined on a field of sets $\mathcal{A}$ such that·

$$\mu(A) \in [0, \infty] \text{ whenever } A \in \mathcal{A}.\,.\, \mu(\varnothing) = 0.\,.$$

$$\mu(A_1 \cup A_2) = \mu(A_1) + \mu(A_2) \text{ whenever } A_1, A_2 \in \mathcal{A} \text{ and } A_1 \cap A_2 = \varnothing.$$

   A very important type of _____ is a measure, which is a σ-additive _____ defined on a σ-field. Every measure is a _____, but not vice-versa.

   a. Content
   b. Dynkin system
   c. Finite character
   d. Finite intersection property

3. _____, Programme and Project Management Maturity Model is a reference guide for structured best practice. It breaks down the broad disciplines of portfolio, programme and project management into a hierarchy of Key Process Areas (KPAs). The hierarchical approach enables organisations to assess their current capability and then plot a roadmap for improvement prioritised by those KPAs which will make the biggest impact on performance.

   a. Phased implementation
   b. PM Declaration of Interdependence
   c. Pmhub
   d. P3M3

Chapter 5. Writing Business Messages

4. Normally, the _____ of a country consists of everyone of working age (typically above a certain age (around 14 to 16) and below retirement (around 65) who are participating workers, that is people actively employed or seeking employment. People not counted include students, retired people, stay-at-home parents, people in prisons or similar institutions, people employed in jobs or professions with unreported income, as well as discouraged workers who cannot find work. _____ in the United States

   In the United States, the unemployment rate is estimated by a household survey called the Current Population Survey, conducted monthly by the Federal Bureau of Labor Statistics.

   a. Labor mobility
   b. Labor shortage
   c. prevailing wage
   d. Labor force

5. _____ is a branch of engineering dealing with the use of computers and telecommunications equipment to store, retrieve, transmit and manipulate data. The term is commonly used as a synonym for computers and computer networks, but it also encompasses other information distribution technologies such as television and telephones. Some of the modern and emerging fields of _____ are next generation web technologies, bioinformatics, cloud computing, global information systems, large-scale knowledge bases, etc.

   a. Information technology
   b. Edith of Wilton
   c. Alain Chartier
   d. Commons-based peer production

1. d
2. a
3. d
4. d
5. a

You can take the complete Chapter Practice Test

for Chapter 5. Writing Business Messages
on all key terms, persons, places, and concepts.

Online 99 Cents

http://www.epub2.4.21191.5.cram101.com/

Use www.Cram101.com for all your study needs

including Cram101's online interactive problem solving labs in

chemistry, statistics, mathematics, and more.

Content

Cultural identity

Formality

Intercultural

Intercultural communication

Social networking

Discovery

Storyboard

Dangling modifier

Information technology

Social media

Consumer

Request for production

Statement

Collaborative workspace

Thesaurus

Job interview

Production

Element

Marking out

Page layout

Virtual team

Cell phone

White Space

Helvetica

Compatibility

Information architecture

Video clip

Desktop publishing

Header

Proposal

Online shopping

Sales pitch

Content	In mathematics, a content is a real function μ defined on a field of sets $\mathcal{A}$ such that· $$\mu(A) \in [0, \infty] \text{ whenever } A \in \mathcal{A}.. \mu(\varnothing) = 0..$$ $\mu(A_1 \cup A_2) = \mu(A_1) + \mu(A_2)$ whenever $A_1, A_2 \in \mathcal{A}$ and $A_1 \cap A_2 = \varnothing$. A very important type of content is a measure, which is a σ-additive content defined on a σ-field. Every measure is a content, but not vice-versa.
Cultural identity	Cultural identity is the identity of a group or culture, or of an individual as far as one is influenced by one's belonging to a group or culture. Cultural identity is similar to and has overlaps with, but is not synonymous with, identity politics. Description Various modern cultural studies and social theories have investigated cultural identity.
Formality	Utterances, conceptually similar to a ritual although typically secular and less involved. A formality may be as simple as a handshake upon making new acquaintances in Western culture to the carefully defined procedure of bows, handshakes, formal greetings, and business-card exchanges that may mark two businessmen being introduced in Japan. In legal and diplomatic circles, formalities include such matters as greeting an arriving head of state with the appropriate national anthem.
Intercultural	Cross-cultural communication (also frequently referred to as intercultural communication, which is also used in a different sense, though) is a field of study that looks at how people from differing cultural backgrounds communicate, in similar and different ways among themselves, and how they endeavour to communicate across cultures. Origins During the Cold War, the United States economy was largely self-contained because the world was polarized into two separate and competing powers: the east and west. However, changes and advancements in economic relationships, political systems, and technological options began to break down old cultural barriers.
Intercultural communication	Intercultural communication is a form of global communication. It is used to describe the wide range of communication problems that naturally appear within an organization made up of individuals from different religious, social, ethnic, and educational backgrounds. Intercultural communication is sometimes used synonymously with cross-cultural communication.
Social networking	A social networking service is an online service, platform, or site that focuses on facilitating the building of social networks or social relations among people who, for example, share interests, activities, backgrounds, or real-life connections. A social network service consists of a representation of each user (often a profile), his/her social links, and a variety of additional services.

Chapter 6. Completing Business Messages

Discovery	Discovery is the act of detecting something new, or something 'old' that had been unknown. With reference to science and academic disciplines, discovery is the observation of new phenomena, new actions, or new events and providing new reasoning to explain the knowledge gathered through such observations with previously acquired knowledge from abstract thought and everyday experiences. Visual discoveries are often called sightings.
Storyboard	Storyboards are graphic organizers in the form of illustrations or images displayed in sequence for the purpose of pre-visualizing a motion picture, animation, motion graphic or interactive media sequence. The storyboarding process, in the form it is known today, was developed at the Walt Disney Studio during the early 1930s, after several years of similar processes being in use at Walt Disney and other animation studios. Origins The storyboarding process can be very time-consuming and intricate.
Dangling modifier	A dangling modifier is an ambiguous grammatical construct, often considered an error in prescriptivist accounts of English, whereby a grammatical modifier could be misinterpreted as being associated with a word other than the one intended, or with no particular word at all. For example, a writer may have meant to modify the subject, but word order makes the modifier seem to modify an object instead. Such ambiguities can lead to unintentional humor or difficulty in understanding a sentence in formal contexts.
Information technology	Information technology is a branch of engineering dealing with the use of computers and telecommunications equipment to store, retrieve, transmit and manipulate data. The term is commonly used as a synonym for computers and computer networks, but it also encompasses other information distribution technologies such as television and telephones. Some of the modern and emerging fields of Information technology are next generation web technologies, bioinformatics, cloud computing, global information systems, large-scale knowledge bases, etc.
Social media	Social media includes web- and mobile-based technologies which are used to turn communication into interactive dialogue among organizations, communities, and individuals. Andreas Kaplan and Michael Haenlein define social media as 'a group of Internet-based applications that build on the ideological and technological foundations of Web 2.0, and that allow the creation and exchange of user-generated content.' When the technologies are in place, social media is ubiquitously accessible, and enabled by scalable communication techniques. Social media technologies take on many different forms including magazines, Internet forums, weblogs, social blogs, microblogging, wikis, social networks, podcasts, photographs or pictures, video, rating and social bookmarking.

Consumer	A consumer is a person or group of people that are the final users of products and or services generated within a social system. A consumer may be a person or group, such as a household. The concept of a consumer may vary significantly by context.
Request for production	A request for production is a legal request for documents, electronically stored information, or other tangible items. In civil procedure, during the discovery phase of litigation, a party to a lawsuit may request that another party provide any documents that it has that pertain to the subject matter of the lawsuit. For example, a party in a court case may obtain copies of e-mail messages sent by employees of the opposing party.
Statement	In logic a statement is either (a) a meaningful declarative sentence that is either true or false, or (b) what is asserted or made by the use of a declarative sentence. In the latter case, a statement is distinct from a sentence in that a sentence is only one formulation of a statement, whereas there may be many other formulations expressing the same statement. Philosopher of language, Peter Strawson advocated the use of the term 'statement' in sense (b) in preference to proposition.
Collaborative workspace	A collaborative workspace is an inter-connected environment in which all the participants in dispersed locations can access and interact with each other just as inside a single entity. The environment may be supported by electronic communications and groupware which enable participants to overcome space and time differentials. These are typically enabled by a shared mental model, common information, and a shared understanding by all of the participants regardless of physical location.
Thesaurus	A thesaurus is a reference work that lists words grouped together according to similarity of meaning , in contrast to a dictionary, which contains definitions and pronunciations. The largest thesaurus in the world is the Historical Thesaurus of the Oxford English Dictionary, which contains more than 920,000 entries. In antiquity, Philo of Byblos authored the first text that could now be called a thesaurus.
Job interview	A job interview is a process in which a potential employee is evaluated by an employer for prospective employment in their company, organization, or firm. During this process, the employer hopes to determine whether or not the applicant is suitable for the role. A job interview typically precedes the hiring decision, and is used to evaluate the candidate.
Production	In economics, production is the act of creating output, a good or service which has value and contributes to the utility of individuals.

Chapter 6. Completing Business Messages

The act may or may not include factors of production other than labor. Any effort directed toward the realization of a desired product or service is a 'productive' effort and the performance of such act is production.

Element	Under United States law, an element of a crime is one of a set of facts that must all be proven to convict a defendant of a crime. Before a court finds a defendant guilty of a criminal offense, the prosecution must present evidence that, even when opposed by any evidence the defense may choose to present, is credible and sufficient to prove beyond a reasonable doubt that the defendant committed each element of the particular crime charged. The component parts that make up any particular crime vary depending on the crime.
Marking out	Marking out is the process of transferring a design or pattern to a workpiece, as the first step in the manufacturing process. It is performed in many industries or hobbies although in the repetition industries the machine's initial setup is designed to remove the need to mark out every individual piece. Marking out consists of transferring the dimensions from the plan to the workpiece in preparation for the next step, machining or manufacture.
Page layout	Page layout is the part of graphic design that deals in the arrangement and style treatment of elements (content) on a page. Beginning from early illuminated pages in hand-copied books of the Middle Ages and proceeding down to intricate modern magazine and catalog layouts, proper page design has long been a consideration in printed material. With print media, elements usually consist of type (text), images (pictures), and occasionally place-holder graphics for elements that are not printed with ink such as dielaser cutting, foil stamping or blind embossing.
Virtual team	A virtual team is a group of individuals who work across time, space and organizational boundaries with links strengthened by webs of communication technology' Ale Ebrahim, N., Ahmed, S. & Taha, Z. in their recent (2009) literature review paper, added two key issues to definition of a virtual team 'as small temporary groups of geographically, organizationally and/ or time dispersed knowledge workers who coordinate their work predominantly with electronic information and communication technologies in order to accomplish one or more organization tasks' Members of virtual teams communicate electronically and may never meet face-to-face. Virtual teams are made possible by a proliferation of fiber optic technology that has significantly increased the scope of off-site communication.
Cell phone	A mobile phone (also known as a cellular phone, cell phone and a hand phone) is a device that can make and receive telephone calls over a radio link while moving around a wide geographic area.

	It does so by connecting to a cellular network provided by a mobile phone operator, allowing access to the public telephone network. By contrast, a cordless telephone is used only within the short range of a single, private base station.
White Space	In process management, the White Space as described by Geary A. Rummler and Alan P. Brache in 1991, is the area between the boxes in an organization chart or the area between the different functions: Very often no one is in charge or responsible for the White Space. The important handoffs between functions are happening here, and this is very often the area where an organization has the greatest potential for improvements. In the White Space things often 'fall between the cracks' or 'disappear into black holes', resulting in misunderstandings and delays.
Helvetica	Helvetica is an independent feature-length documentary film about typography and graphic design, centered on the typeface of the same name. Directed by Gary Hustwit, it was released in 2007 to coincide with the 50th anniversary of the typeface's introduction in 1957 and is considered the first of the Design Trilogy by the director. Its content consists of a history of the typeface interspersed with candid interviews with leading graphic and type designers.
Compatibility	Chemical compatibility is a measure of how stable a substance is when mixed with another substance. If substances mix and do not change they are considered compatible. If substances mix and change or do not mix at all they are considered incompatible.
Information architecture	Information architecture is the art and science of organizing and labelling websites, intranets, online communities and software to support usability. It is an emerging discipline and community of practice focused on bringing together principles of design and architecture to the digital landscape. Typically it involves a model or concept of information which is used and applied to activities that require explicit details of complex information systems.
Video clip	Video clips are short clips of video, usually part of a longer recording. The term is also more loosely used to mean any short video less than the length of a traditional television program. On the Internet With the spread of Internet global accessing(fastest Internet broadband connection of TCP with accumulator cables and semi fast connection), video clips have become very popular online.
Desktop publishing	Desktop publishing is the creation of printed materials using page layout software on a personal computer. When used skilfully, desktop publishing can produce printed literature with attractive layouts and typographic quality comparable to traditional typography and printing.

Chapter 6. Completing Business Messages

Header	In information technology, header refers to supplemental data placed at the beginning of a block of data being stored or transmitted. In data transmission, the data following the header are sometimes called the payload or body. It is vital that header composition follow a clear and unambiguous specification or format, to allow for parsing.
Proposal	A business proposal is a written offer from a seller to a prospective buyer. Business proposals are often a key step in the complex sales process--i.e., whenever a buyer considers more than price in a purchase. There are three distinct categories of business proposals:•formally solicited•informally solicited•unsolicited. Solicited proposals are written in response to published requirements, contained in a Request for Proposal (RFP), Request for Quotation (RFQ), Request for Information (RFI) or an Invitation For Bid (IFB).
Online shopping	Online shopping is a form of electronic commerce whereby consumers directly buy goods or services from a seller over the Internet without an intermediary service. An online shop, eshop, e-store, Internet shop, webshop, webstore, online store, or virtual store evokes the physical analogy of buying products or services at a bricks-and-mortar retailer or shopping centre. The process is called business-to-consumer (B2C) online shopping.
Sales pitch	In selling technique, a sales pitch is a line of talk that attempts to persuade someone or something, with a planned sales presentation strategy of a product or service designed to initiate and close a sale of the product or service. A sales pitch is a planned presentation of a product or service designed to initiate and close a sale of the same product or service. A sales pitch is essentially designed to be either an introduction of a product or service to an audience who knows nothing about it, or a descriptive expansion of a product or service that an audience has already expressed interest in.

1. Utterances, conceptually similar to a ritual although typically secular and less involved. A _____ may be as simple as a handshake upon making new acquaintances in Western culture to the carefully defined procedure of bows, handshakes, formal greetings, and business-card exchanges that may mark two businessmen being introduced in Japan. In legal and diplomatic circles, _____(ies) include such matters as greeting an arriving head of state with the appropriate national anthem.

   a. Jimmy Carter
   b. Formality
   c. Finite character
   d. Finite intersection property

2. In mathematics, a _____ is a real function μ defined on a field of sets $\mathcal{A}$ such that•

$$\mu(A) \in [0, \infty] \text{ whenever } A \in \mathcal{A}., \mu(\varnothing) = 0.,$$
$$\mu(A_1 \cup A_2) = \mu(A_1) + \mu(A_2) \text{ whenever } A_1, A_2 \in \mathcal{A} \text{ and } A_1 \cap A_2 = \varnothing.$$

   A very important type of _____ is a measure, which is a σ-additive _____ defined on a σ-field. Every measure is a _____, but not vice-versa.

   a. Content
   b. Dynkin system
   c. Finite character
   d. Finite intersection property

3. _____ is the identity of a group or culture, or of an individual as far as one is influenced by one's belonging to a group or culture. _____ is similar to and has overlaps with, but is not synonymous with, identity politics. Description

   Various modern cultural studies and social theories have investigated _____.

   a. Jimmy Carter
   b. Dynkin system
   c. Finite character
   d. Cultural identity

4. A mobile phone (also known as a cellular phone, _____ and a hand phone) is a device that can make and receive telephone calls over a radio link while moving around a wide geographic area. It does so by connecting to a cellular network provided by a mobile phone operator, allowing access to the public telephone network. By contrast, a cordless telephone is used only within the short range of a single, private base station.

   a. Jimmy Carter
   b. Multifactor
   c. Simple interest
   d. Cell phone

Chapter 6. Completing Business Messages

5. _____ is a branch of engineering dealing with the use of computers and telecommunications equipment to store, retrieve, transmit and manipulate data. The term is commonly used as a synonym for computers and computer networks, but it also encompasses other information distribution technologies such as television and telephones. Some of the modern and emerging fields of _____ are next generation web technologies, bioinformatics, cloud computing, global information systems, large-scale knowledge bases, etc.

    a. Offshore outsourcing
    b. Edith of Wilton
    c. Information technology
    d. Habituation

ANSWER KEY
Chapter 6. Completing Business Messages

---

1. b
2. a
3. d
4. d
5. c

You can take the complete Chapter Practice Test

for Chapter 6. Completing Business Messages
on all key terms, persons, places, and concepts.

Online 99 Cents

http://www.epub2.4.21191.6.cram101.com/

Use www.Cram101.com for all your study needs

including Cram101's online interactive problem solving labs in

chemistry, statistics, mathematics, and more.

Lie-Nielsen Toolworks

YouTube

Business communication

Cultural identity

Personal branding

Social media

Baby boomer

Microblogging

Podcasts

Proposal

Social bookmarking

Social network

User-generated content

Financial Services

Small business

Barcode

Orientation

Social networking

Intercultural

Intercultural communication

Marketing Intelligence

Storyboard

Content

Request for production

Lawsuit

Misinformation

Sales pitch

Word-of-mouth marketing

Search Engine

Market research

Marketing

Public opinion

P3M3

Quizzle

Brand community

Project management

Writing style

Group buying

CHAPTER OUTLINE: KEY TERMS, PEOPLE, PLACES, CONCEPTS

Consumer

Crowdsourcing

Information product

Virtual team

Formality

Online shopping

Nutrition

Customer support

Topics

Public relations

Entrepreneurial culture

Viral marketing

Credibility

Hashtag

Chapter 7. Crafting Messages for Electronic Media

Lie-Nielsen Toolworks	Lie-Nielsen Toolworks, Inc. is a family-owned business, established in 1981 and based in Warren, Maine. It manufactures a range of hand tools, primarily for woodworking, based on traditional designs.
YouTube	YouTube is a video-sharing website, created by three former PayPal employees in February 2005, on which users can upload, view and share videos. The company is based in San Bruno, California, and uses Adobe Flash Video and HTML5 technology to display a wide variety of user-generated video content, including movie clips, TV clips, and music videos, as well as amateur content such as video blogging, short original videos, and educational videos. Most of the content on YouTube has been uploaded by individuals, although media corporations including CBS, the BBC, VEVO, Hulu, and other organizations offer some of their material via the site, as part of the YouTube partnership program.
Business communication	Business Communication: communication used to promote a product, service, or organization; relay information within the business; or deal with legal and similar issues. It is also a means of relaying between a supply chain, for example the consumer and manufacturer. Business Communication is known simply as 'communications'.
Cultural identity	Cultural identity is the identity of a group or culture, or of an individual as far as one is influenced by one's belonging to a group or culture. Cultural identity is similar to and has overlaps with, but is not synonymous with, identity politics. Description Various modern cultural studies and social theories have investigated cultural identity.
Personal branding	Personal branding is, for some people, a description of the process whereby people and their careers are marked as brands. It has been noted that while previous self-help management techniques were about self-improvement, the personal branding concept suggests instead that success comes from self-packaging. Further defined as the creation of an asset that pertains to a particular person or individual; this includes but is not limited to the body, clothing, appearance and knowledge contained within, leading to an indelible impression that is uniquely distinguishable.
Social media	Social media includes web- and mobile-based technologies which are used to turn communication into interactive dialogue among organizations, communities, and individuals. Andreas Kaplan and Michael Haenlein define social media as 'a group of Internet-based applications that build on the ideological and technological foundations of Web 2.0, and that allow the creation and exchange of user-generated content.' When the technologies are in place, social media is ubiquitously accessible, and enabled by scalable communication techniques.

Baby boomer	A baby boomer is a person who was born during the demographic Post-World War II baby boom between the years 1946 and 1964, according to the U.S. Census Bureau. The term 'baby boomer' is sometimes used in a cultural context. Therefore, it is impossible to achieve broad consensus of a precise definition, even within a given territory.
Microblogging	Microblogging is a broadcast medium in the form of blogging. A microblog differs from a traditional blog in that its content is typically smaller in both actual and aggregate file size. Microblogs 'allow users to exchange small elements of content such as short sentences, individual images, or video links'.
Podcasts	A podcast is a type of digital media consisting of an episodic series of audio radio, video, PDF, or ePub files subscribed to and downloaded through web syndication or streamed online to a computer or mobile device. The word is a neologism derived from 'broadcast' and 'pod' from the success of the iPod, as podcasts are often listened to on portable media players. In the context of Apple devices, the term 'Podcasts' refers to the audio and video version of podcasts, whereas the textual version of podcasts are classified under the app known as Newsstand.
Proposal	A business proposal is a written offer from a seller to a prospective buyer. Business proposals are often a key step in the complex sales process--i.e., whenever a buyer considers more than price in a purchase. There are three distinct categories of business proposals:•formally solicited•informally solicited•unsolicited. Solicited proposals are written in response to published requirements, contained in a Request for Proposal (RFP), Request for Quotation (RFQ), Request for Information (RFI) or an Invitation For Bid (IFB).
Social bookmarking	Social bookmarking is a method for Internet users to organize, store, manage and search for bookmarks of resources online. Many online bookmark management services have launched since 1996; Delicious, founded in 2003, popularized the terms 'social bookmarking' and 'tagging'. Tagging is a significant feature of social bookmarking systems, enabling users to organize their bookmarks in flexible ways and develop shared vocabularies known as folksonomies.
Social network	A social network is a social structure made up of a set of actors (such as individuals or organizations) and the dyadic ties between these actors. The social network perspective provides a clear way of analyzing the structure of whole social entities. The study of these structures uses social network analysis to identify local and global patterns, locate influential entities, and examine network dynamics.

Chapter 7. Crafting Messages for Electronic Media

User-generated content	User-generated content covers a range of media content available in a range of modern communications technologies. It entered mainstream usage during 2005, having arisen in web publishing and new media content production circles. Its use for a wide range of applications, including problem processing, news, gossip and research, reflects the expansion of media production through new technologies that are accessible and affordable to the general public.
Financial Services	Financial services are the economic services provided by the finance industry, which encompasses a broad range of organizations that manage money, including credit unions, banks, credit card companies, insurance companies, consumer finance companies, stock brokerages, investment funds and some government sponsored enterprises. As of 2004, the financial services industry represented 20% of the market capitalization of the S&P 500 in the United States. History of financial services

The term 'financial services' became more prevalent in the United States partly as a result of the Gramm-Leach-Bliley Act of the late 1990s, which enabled different types of companies operating in the U.S. financial services industry at that time to merge. |
Small business	What constitutes a small business varies widely around the world. Small businesses are normally privately owned corporations, partnerships, or sole proprietorships. What constitutes 'small' in terms of government support and tax policy varies by country and by industry, ranging from fewer than 15 employees under the Australian Fair Work Act 2009, 50 employees according to the definition used by the European Union, and fewer than 500 employees to qualify for many U.S. Small Business Administration programs, although in 2006 there were over 18,000 'small businesses' with over 500 employees that accounted for half of all the employees employed by all 'small business '.
Barcode	A barcode is an optical machine-readable representation of data, which shows data about the object to which it attaches. Originally barcodes represented data by varying the widths and spacings of parallel lines, and may be referred to as linear or one-dimensional (1D). Later they evolved into rectangles, dots, hexagons and other geometric patterns in two dimensions (2D).
Orientation	In mathematics, orientation is a geometric notion that in two dimensions allows one to say when a cycle goes around clockwise or counterclockwise, and in three dimensions when a figure is left-handed or right-handed. In linear algebra, the notion of orientation makes sense in arbitrary dimensions. In this setting, the orientation of an ordered basis is a kind of asymmetry that makes a reflection impossible to replicate by means of a simple rotation.
Social networking	A social networking service is an online service, platform, or site that focuses on facilitating the building of social networks or social relations among people who, for example, share interests, activities, backgrounds, or real-life connections.

	A social network service consists of a representation of each user (often a profile), his/her social links, and a variety of additional services. Most social network services are web-based and provide means for users to interact over the Internet, such as e-mail and instant messaging.
Intercultural	Cross-cultural communication (also frequently referred to as intercultural communication, which is also used in a different sense, though) is a field of study that looks at how people from differing cultural backgrounds communicate, in similar and different ways among themselves, and how they endeavour to communicate across cultures. Origins

During the Cold War, the United States economy was largely self-contained because the world was polarized into two separate and competing powers: the east and west. However, changes and advancements in economic relationships, political systems, and technological options began to break down old cultural barriers. |
| Intercultural communication | Intercultural communication is a form of global communication. It is used to describe the wide range of communication problems that naturally appear within an organization made up of individuals from different religious, social, ethnic, and educational backgrounds. Intercultural communication is sometimes used synonymously with cross-cultural communication. |
| Marketing Intelligence | Marketing Intelligence is the information relevant to a company's markets, gathered and analyzed specifically for the purpose of accurate and confident decision-making in determining market opportunity, market penetration strategy, and market development metrics. Marketing intelligence is necessary when entering a foreign market. Marketing Intelligence is not the same as Market Intelligence (MARKINT). |
| Storyboard | Storyboards are graphic organizers in the form of illustrations or images displayed in sequence for the purpose of pre-visualizing a motion picture, animation, motion graphic or interactive media sequence.

The storyboarding process, in the form it is known today, was developed at the Walt Disney Studio during the early 1930s, after several years of similar processes being in use at Walt Disney and other animation studios. Origins

The storyboarding process can be very time-consuming and intricate. |
| Content | In mathematics, a content is a real function μ defined on a field of sets $\mathcal{A}$ such that
$$\mu(A) \in [0, \infty] \text{ whenever } A \in \mathcal{A}. \, \mu(\varnothing) = 0.$$
$\mu(A_1 \cup A_2) = \mu(A_1) + \mu(A_2)$ whenever $A_1, A_2 \in \mathcal{A}$ and $A_1 \cap A_2 = \varnothing$.

A very important type of content is a measure, which is a σ-additive content defined on a σ-field. |

Chapter 7. Crafting Messages for Electronic Media

Request for production	A request for production is a legal request for documents, electronically stored information, or other tangible items. In civil procedure, during the discovery phase of litigation, a party to a lawsuit may request that another party provide any documents that it has that pertain to the subject matter of the lawsuit. For example, a party in a court case may obtain copies of e-mail messages sent by employees of the opposing party.
Lawsuit	A lawsuit is a civil action brought in a court of law in which a plaintiff, a party who claims to have incurred loss as a result of a defendant's actions, demands a legal or equitable remedy. The defendant is required to respond to the plaintiff's complaint. If the plaintiff is successful, judgment will be given in the plaintiff's favor, and a variety of court orders may be issued to enforce a right, award damages, or impose a temporary or permanent injunction to prevent an act or compel an act.
Misinformation	Misinformation is false or inaccurate information that is spread unintentionally. It is distinguished from disinformation by motive in that misinformation is simply erroneous, while disinformation, in contrast, is intended to mislead. Adam Makkai proposes the distinction between misinformation and disinformation to be a defining characteristic of idioms in the English language.
Sales pitch	In selling technique, a sales pitch is a line of talk that attempts to persuade someone or something, with a planned sales presentation strategy of a product or service designed to initiate and close a sale of the product or service. A sales pitch is a planned presentation of a product or service designed to initiate and close a sale of the same product or service. A sales pitch is essentially designed to be either an introduction of a product or service to an audience who knows nothing about it, or a descriptive expansion of a product or service that an audience has already expressed interest in.
Word-of-mouth marketing	Word-of-mouth marketing also called word of mouth advertising, is an unpaid form of promotion--oral or written--in which satisfied customers tell other people how much they like a business, product, service, or event. Word-of-mouth is one of the most credible forms of advertising because people who don't stand to gain personally by promoting something put their reputations on the line every time they make a recommendation. George Silverman, a psychologist, pioneered word-of-mouth marketing when he created what he called 'teleconferenced peer influence groups' in order to engage physicians in dialogue about new pharmaceutical products.
Search Engine	Search Engine was a weekly Canadian radio show that aired on CBC Radio One, then a dedicated podcast available on CBC.ca, and now a podcast on TVOntario's website, tvo.org.

	It is hosted by Jesse Brown, who also co-produces the show with Geoff Siskind and Andrew Parker. Cory Doctorow, novelist and editor of Boing Boing, is also a regular contributor.
Market research	Market research is any organized effort to gather information about markets or customers. It is a very important component of business strategy. The term is commonly interchanged with marketing research; however, expert practitioners may wish to draw a distinction, in that marketing research is concerned specifically about marketing processes, while market research is concerned specifically with markets.
Marketing	Marketing is 'the activity, set of institutions, and processes for creating, communicating, delivering, and exchanging offerings that have value for customers, clients, partners, and society at large.' For business to consumer marketing, it is 'the process by which companies create value for customers and build strong customer relationships, in order to capture value from customers in return'. For business to business marketing it is creating value, solutions, and relationships either short term or long term with a company or brand. It generates the strategy that underlies sales techniques, business communication, and business developments.
Public opinion	Public opinion is the aggregate of individual attitudes or beliefs held by the adult population. Public opinion can also be defined as the complex collection of opinions of many different people and the sum of all their views. The principle approaches to the study of public opinion may be divided into 4 categories:•quantitative measurement of opinion distributions;•investigation of the internal relationships among the individual opinions that make up public opinion on an issue;•description or analysis of the public role of public opinion;•study both of the communication media that disseminate the ideas on which opinions are based and of the uses that propagandists and other manipulators make of these media.Concepts of public opinion Public opinion as a concept gained credence with the rise of 'public' in the eighteenth century.
P3M3	P3M3, Programme and Project Management Maturity Model is a reference guide for structured best practice. It breaks down the broad disciplines of portfolio, programme and project management into a hierarchy of Key Process Areas (KPAs). The hierarchical approach enables organisations to assess their current capability and then plot a roadmap for improvement prioritised by those KPAs which will make the biggest impact on performance.
Quizzle	Quizzle.com, a Detroit-based website, is the first company to give consumers free access to their credit score. Quizzle provides consumers with CE credit score based on an Experian credit report.

Chapter 7. Crafting Messages for Electronic Media

Brand community	A brand community is a community formed on the basis of attachment to a product or marque. Recent developments in marketing and in research in consumer behavior result in stressing the connection between brand, individual identity and culture. Among the concepts developed to explain the behavior of consumers, the concept of a brand community focuses on the connections between consumers.
Project management	Project management is the discipline of planning, organizing, securing, managing, leading, and controlling resources to achieve specific goals. A project is a temporary endeavor with a defined beginning and end (usually time-constrained, and often constrained by funding or deliverables), undertaken to meet unique goals and objectives, typically to bring about beneficial change or added value. The temporary nature of projects stands in contrast with business as usual , which are repetitive, permanent, or semi-permanent functional activities to produce products or services.
Writing style	Writing style is the manner in which an author chooses to write to his or her audience. A style reveals both the writer's personality and voice, but it also shows how she or he perceives the audience, and chooses conceptual writing style which reveal those choices by which the writer may change the conceptual world of the overall character of the work. This might be done by a simple change of words; a syntactical structure, parsing prose, adding diction, and organizing figures of thought into usable frameworks.
Group buying	Group buying, offers products and services at significantly reduced prices on the condition that a minimum number of buyers would make the purchase. Origins of group buying can be traced to China where tuángòu or team buying was executed to get discount prices from retailer when a large group of people were willing to buy the same item. In recent time, group buying websites have emerged as a major player in online shopping business.
Consumer	A consumer is a person or group of people that are the final users of products and or services generated within a social system. A consumer may be a person or group, such as a household. The concept of a consumer may vary significantly by context.
Crowdsourcing	Crowdsourcing is a process that involves outsourcing tasks to a distributed group of people. This process can occur both online and offline. The difference between crowdsourcing and ordinary outsourcing is that a task or problem is outsourced to an undefined public rather than a specific body, such as paid employees.
Information product	An Information Product is any final product in the form of information that a person needs to have. This Information Product consists of several Information Element, which are located in the organizational value chain. To illustrate the concept of an IP, an example is shown of a bottleneck analysis in HR (by J. Willems 2008).

Virtual team	A virtual team is a group of individuals who work across time, space and organizational boundaries with links strengthened by webs of communication technology' Ale Ebrahim, N., Ahmed, S. & Taha, Z. in their recent (2009) literature review paper, added two key issues to definition of a virtual team 'as small temporary groups of geographically, organizationally and/ or time dispersed knowledge workers who coordinate their work predominantly with electronic information and communication technologies in order to accomplish one or more organization tasks' Members of virtual teams communicate electronically and may never meet face-to-face. Virtual teams are made possible by a proliferation of fiber optic technology that has significantly increased the scope of off-site communication.
Formality	Utterances, conceptually similar to a ritual although typically secular and less involved. A formality may be as simple as a handshake upon making new acquaintances in Western culture to the carefully defined procedure of bows, handshakes, formal greetings, and business-card exchanges that may mark two businessmen being introduced in Japan. In legal and diplomatic circles, formalities include such matters as greeting an arriving head of state with the appropriate national anthem.
Online shopping	Online shopping is a form of electronic commerce whereby consumers directly buy goods or services from a seller over the Internet without an intermediary service. An online shop, eshop, e-store, Internet shop, webshop, webstore, online store, or virtual store evokes the physical analogy of buying products or services at a bricks-and-mortar retailer or shopping centre. The process is called business-to-consumer (B2C) online shopping.
Nutrition	Nutrition is the provision, to cells and organisms, of the materials necessary (in the form of food) to support life. Many common health problems can be prevented or alleviated with a healthy diet. The diet of an organism is what it eats, which is largely determined by the perceived palatability of foods.
Customer support	Customer support is a range of customer services to assist customers in making cost effective and correct use of a product. It includes assistance in planning, installation, training, trouble shooting, maintenance, upgrading, and disposal of a product. Regarding technology products such as mobile phones, televisions, computers, software products or other electronic or mechanical goods, it is termed technical support.
Topics	The Topics is the name given to one of Aristotle's six works on logic collectively known as the Organon. The other five are:•Categories•De Interpretatione•Prior Analytics•Posterior Analytics•On Sophistical Refutations

Chapter 7. Crafting Messages for Electronic Media

	The Topics constitutes Aristotle's treatise on the art of dialectic--the invention and discovery of arguments in which the propositions rest upon commonly-held opinions or endoxa . Topoi are 'places' from which such arguments can be discovered or invented.
Public relations	Public relations is the practice of managing the flow of information between an individual or an organization and the public. Public relations provides an organization or individual exposure to their audiences using topics of public interest and news items that do not require direct payment. The aim of public relations by a company often is to persuade the public, investors, partners, employees, and other stakeholders to maintain a certain point of view about it, its leadership, products, or of political decisions.
Entrepreneurial culture	Entrepreneurial culture, is a form of ideal, which is based on the value system of an enterprise and closely related to the management philosophy as well as the management behaviour of the enterprise. It is where the kernel of business management lies. In the narrow sense, it refers to some fundamental spirit and agglomerating force that come into being in the production and management practices of an enterprise, as well as the common values and norms of behavior shared by the whole staff.
Viral marketing	Viral marketing, viral advertising, or marketing buzz are buzzwords referring to marketing techniques that use pre-existing social networks to produce increases in brand awareness or to achieve other marketing objectives (such as product sales) through self-replicating viral processes, analogous to the spread of viruses or computer viruses (cf. internet memes and memetics). It can be delivered by word of mouth or enhanced by the network effects of the Internet.
Credibility	Credibility refers to the objective and subjective components of the believability of a source or message. Traditionally, modern, credibility has two key components: trustworthiness and expertise, which both have objective and subjective components. Trustworthiness is based more on subjective factors, but can include objective measurements such as established reliability.
Hashtag	Hashtags are words or phrases prefixed with the symbol , a form of metadata tag. They are used within IRC networks to identify groups and topics. Also, short messages on microblogging social networking services such as Twitter, identi.ca or Google+ may be tagged by including one or more with multiple words concatenated, e.g.:#Wikipedia is an #encyclopedia Searching for the string #Wikipedia will cause this word to appear in the search engine results.

1. _____ is, for some people, a description of the process whereby people and their careers are marked as brands. It has been noted that while previous self-help management techniques were about self-improvement, the _____ concept suggests instead that success comes from self-packaging. Further defined as the creation of an asset that pertains to a particular person or individual; this includes but is not limited to the body, clothing, appearance and knowledge contained within, leading to an indelible impression that is uniquely distinguishable.

   a. Personal identity
   b. Self-perception theory
   c. Self-schema
   d. Personal branding

2. _____ includes web- and mobile-based technologies which are used to turn communication into interactive dialogue among organizations, communities, and individuals. Andreas Kaplan and Michael Haenlein define _____ as 'a group of Internet-based applications that build on the ideological and technological foundations of Web 2.0, and that allow the creation and exchange of user-generated content.' When the technologies are in place, _____ is ubiquitously accessible, and enabled by scalable communication techniques. Classification of _____

   _____ technologies take on many different forms including magazines, Internet forums, weblogs, social blogs, microblogging, wikis, social networks, podcasts, photographs or pictures, video, rating and social bookmarking.

   a. Jimmy Carter
   b. Self-perception theory
   c. Self-schema
   d. Social media

3. _____: communication used to promote a product, service, or organization; relay information within the business; or deal with legal and similar issues. It is also a means of relaying between a supply chain, for example the consumer and manufacturer.

   _____ is known simply as 'communications'.

   a. Business logic
   b. Business communication
   c. Business Process Definition Metamodel
   d. Business process management

4. . Cross-cultural communication (also frequently referred to as _____ communication, which is also used in a different sense, though) is a field of study that looks at how people from differing cultural backgrounds communicate, in similar and different ways among themselves, and how they endeavour to communicate across cultures. Origins

   During the Cold War, the United States economy was largely self-contained because the world was polarized into two separate and competing powers: the east and west. However, changes and advancements in economic relationships, political systems, and technological options began to break down old cultural barriers.

a. Intercultural

b. Orthogonal complement

c. Orthogonal diagonalization

d. Orthogonal Procrustes problem

5. _____, Inc. is a family-owned business, established in 1981 and based in Warren, Maine. It manufactures a range of hand tools, primarily for woodworking, based on traditional designs.

a. Lie-Nielsen Toolworks

b. Negative basis

c. Jimmy Carter

d. Multifactor

1. d

2. d

3. b

4. a

5. a

You can take the complete Chapter Practice Test

for Chapter 7. Crafting Messages for Electronic Media
on all key terms, persons, places, and concepts.

Online 99 Cents

http://www.epub2.4.21191.7.cram101.com/

Use www.Cram101.com for all your study needs

including Cram101's online interactive problem solving labs in

chemistry, statistics, mathematics, and more.

Get Satisfaction

Intercultural

Intercultural communication

Proposal

Search Engine

Golden Rule

Goodwill

Statement

Information technology

Crisis management

Customer satisfaction

Financial Reporting

Letter of recommendation

Cultural identity

Defamation

Social network

Social bookmarking

Job interview

Appreciation

Chapter 8. Writing Routine and Positive Messages

	Brainstorming

Get Satisfaction	Get Satisfaction is a customer community software platform for technical support based in San Francisco, California, United States. It was founded on January 31, 2007 by several people, including Lane Becker, Amy Muller, Thor Muller, and Jonathan Grubb. It publicly launched in September 2007. The idea for the service originated from Valleyschwag as a side project.
Intercultural	Cross-cultural communication (also frequently referred to as intercultural communication, which is also used in a different sense, though) is a field of study that looks at how people from differing cultural backgrounds communicate, in similar and different ways among themselves, and how they endeavour to communicate across cultures. Origins

During the Cold War, the United States economy was largely self-contained because the world was polarized into two separate and competing powers: the east and west. However, changes and advancements in economic relationships, political systems, and technological options began to break down old cultural barriers. |
| Intercultural communication | Intercultural communication is a form of global communication. It is used to describe the wide range of communication problems that naturally appear within an organization made up of individuals from different religious, social, ethnic, and educational backgrounds. Intercultural communication is sometimes used synonymously with cross-cultural communication. |
| Proposal | A business proposal is a written offer from a seller to a prospective buyer. Business proposals are often a key step in the complex sales process--i.e., whenever a buyer considers more than price in a purchase.

There are three distinct categories of business proposals:•formally solicited•informally solicited•unsolicited.

Solicited proposals are written in response to published requirements, contained in a Request for Proposal (RFP), Request for Quotation (RFQ), Request for Information (RFI) or an Invitation For Bid (IFB). |

Search Engine	Search Engine was a weekly Canadian radio show that aired on CBC Radio One, then a dedicated podcast available on CBC.ca, and now a podcast on TVOntario's website, tvo.org. It is hosted by Jesse Brown, who also co-produces the show with Geoff Siskind and Andrew Parker. Cory Doctorow, novelist and editor of Boing Boing, is also a regular contributor.
Golden Rule	The Golden Rule is a maxim, ethical code, or morality that essentially states either of the following:•(Positive form): One should treat others as one would like others to treat oneself.• (Negative/prohibitive form, also called the Silver Rule): One should not treat others in ways that one would not like to be treated.
	This concept describes a 'reciprocal' or 'two-way' relationship between one's self and others that involves both sides equally and in a mutual fashion.
	This concept can be explained from the perspective of psychology, philosophy, sociology, religion, etc.: Psychologically it involves a person empathizing with others. Philosophically it involves a person perceiving their neighbor as also 'an I' or 'self.' Sociologically, this principle is applicable between individuals, between groups, and between individuals and groups.
Goodwill	Goodwill is an accounting concept meaning the value of an entity over and above the value of its assets. The term was originally used in accounting to express the intangible but quantifiable 'prudent value' of an ongoing business beyond its assets, resulting perhaps from the reputation the firm enjoyed with its clients.
	For example, a software company may have net assets (consisting primarily of miscellaneous equipment, and assuming no debt) valued at $1 million, but the company's overall value (including brand, customers, intellectual capital) is valued at $10 million.
Statement	In logic a statement is either (a) a meaningful declarative sentence that is either true or false, or (b) what is asserted or made by the use of a declarative sentence. In the latter case, a statement is distinct from a sentence in that a sentence is only one formulation of a statement, whereas there may be many other formulations expressing the same statement.
	Philosopher of language, Peter Strawson advocated the use of the term 'statement' in sense (b) in preference to proposition.
Information technology	Information technology is a branch of engineering dealing with the use of computers and telecommunications equipment to store, retrieve, transmit and manipulate data. The term is commonly used as a synonym for computers and computer networks, but it also encompasses other information distribution technologies such as television and telephones.

Chapter 8. Writing Routine and Positive Messages

Crisis management	Crisis management is the process by which an organization deals with a major event that threatens to harm the organization, its stakeholders, or the general public. The study of crisis management originated with the large scale industrial and environmental disasters in the 1980s. Three elements are common to most definitions of crisis: (a) a threat to the organization, (b) the element of surprise, and (c) a short decision time.
Customer satisfaction	Customer satisfaction, a term frequently used in marketing, is a measure of how products and services supplied by a company meet or surpass customer expectation. Customer satisfaction is defined as 'the number of customers, or percentage of total customers, whose reported experience with a firm, its products, or its services (ratings) exceeds specified satisfaction goals.' In a survey of nearly 200 senior marketing managers, 71 percent responded that they found a customer satisfaction metric very useful in managing and monitoring their businesses. It is seen as a key performance indicator within business and is often part of a Balanced Scorecard.
Financial Reporting	Financial reporting is the process of preparing and distributing financial information to users of such information in various forms. The most common format of formal financial reporting are financial statements. Financial statements are prepared in accordance with rigorously applied standards defined by professional accounting bodies developed according to the legal and professional framework of a specific locale.
Letter of recommendation	A letter of recommendation is a letter in which the writer assesses the qualities, characteristics, and capabilities of the person being recommended in terms of that individual's ability to perform a particular task or function. Recommendation letters are almost always specifically requested to be written about someone, and are therefore addressed to a particular requestor. Letters of recommendation are typically related to employment, admissions to institutions of higher education or scholarship eligibility.
Cultural identity	Cultural identity is the identity of a group or culture, or of an individual as far as one is influenced by one's belonging to a group or culture. Cultural identity is similar to and has overlaps with, but is not synonymous with, identity politics. Description Various modern cultural studies and social theories have investigated cultural identity.
Defamation	Defamation--also called calumny, vilification, traducement, slander (for transitory statements), and libel (for written, broadcast, or otherwise published words)--is the communication of a statement that makes a claim, expressly stated or implied to be factual, that may give an individual, business, product, group, government, or nation a negative image. This can be also any disparaging statement made by one person about another, which is communicated or published, whether true or false, depending on legal state.

Social network	A social network is a social structure made up of a set of actors (such as individuals or organizations) and the dyadic ties between these actors. The social network perspective provides a clear way of analyzing the structure of whole social entities. The study of these structures uses social network analysis to identify local and global patterns, locate influential entities, and examine network dynamics.
Social bookmarking	Social bookmarking is a method for Internet users to organize, store, manage and search for bookmarks of resources online. Many online bookmark management services have launched since 1996; Delicious, founded in 2003, popularized the terms 'social bookmarking' and 'tagging'. Tagging is a significant feature of social bookmarking systems, enabling users to organize their bookmarks in flexible ways and develop shared vocabularies known as folksonomies.
Job interview	A job interview is a process in which a potential employee is evaluated by an employer for prospective employment in their company, organization, or firm. During this process, the employer hopes to determine whether or not the applicant is suitable for the role. A job interview typically precedes the hiring decision, and is used to evaluate the candidate.
Appreciation	In accounting, appreciation of an asset is an increase in its value. In this sense it is the reverse of depreciation, which measures the fall in value of assets over their normal life-time. Generally, the term is reserved for property or, more specifically, land and buildings.
Brainstorming	Brainstorming is a group or individual creativity technique by which efforts are made to find a conclusion for a specific problem by gathering a list of ideas spontaneously contributed by its member(s). The term was popularized by Alex Faickney Osborn in the 1963 book Applied Imagination. Osborn claimed that brainstorming was more effective than individuals working alone in generating ideas, although more recent research has questioned this conclusion.

Chapter 8. Writing Routine and Positive Messages

1. _____ is the identity of a group or culture, or of an individual as far as one is influenced by one's belonging to a group or culture. _____ is similar to and has overlaps with, but is not synonymous with, identity politics. Description

   Various modern cultural studies and social theories have investigated _____.

   a. Cultural identity
   b. business correspondence
   c. Multifactor
   d. Deferred tax

2. In accounting, _____ of an asset is an increase in its value. In this sense it is the reverse of depreciation, which measures the fall in value of assets over their normal life-time. Generally, the term is reserved for property or, more specifically, land and buildings.

   a. Asset
   b. Appreciation
   c. Internal audit
   d. International Standards on Auditing

3. _____ is an accounting concept meaning the value of an entity over and above the value of its assets. The term was originally used in accounting to express the intangible but quantifiable 'prudent value' of an ongoing business beyond its assets, resulting perhaps from the reputation the firm enjoyed with its clients.

   For example, a software company may have net assets (consisting primarily of miscellaneous equipment, and assuming no debt) valued at $1 million, but the company's overall value (including brand, customers, intellectual capital) is valued at $10 million.

   a. Gross income
   b. Goodwill
   c. Gross profit
   d. Historical cost

4. _____ is a customer community software platform for technical support based in San Francisco, California, United States. It was founded on January 31, 2007 by several people, including Lane Becker, Amy Muller, Thor Muller, and Jonathan Grubb. It publicly launched in September 2007. The idea for the service originated from Valleyschwag as a side project.

   a. Get Satisfaction
   b. GNU Enterprise
   c. GoldMine
   d. Governance, risk management, and compliance

5. Cross-cultural communication (also frequently referred to as _____ communication, which is also used in a different sense, though) is a field of study that looks at how people from differing cultural backgrounds communicate, in similar and different ways among themselves, and how they endeavour to communicate across cultures. Origins

During the Cold War, the United States economy was largely self-contained because the world was polarized into two separate and competing powers: the east and west. However, changes and advancements in economic relationships, political systems, and technological options began to break down old cultural barriers.

a. Intercultural
b. GNU Enterprise
c. GoldMine
d. Governance, risk management, and compliance

1. a
2. b
3. b
4. a
5. a

You can take the complete Chapter Practice Test

for Chapter 8. Writing Routine and Positive Messages
on all key terms, persons, places, and concepts.

Online 99 Cents

http://www.epub2.4.21191.8.cram101.com/

Use www.Cram101.com for all your study needs

including Cram101's online interactive problem solving labs in

chemistry, statistics, mathematics, and more.

Chapter 9. Writing Negative Messages

Public opinion

Search Engine

Independent contractor

Credential

Intercultural

Intercultural communication

Buffer

Job interview

Social network

Podcasts

Cultural identity

Defamation

Information technology

Labor force

Crisis communication

Crisis management

Financial Services

Proposal

Public opinion	Public opinion is the aggregate of individual attitudes or beliefs held by the adult population. Public opinion can also be defined as the complex collection of opinions of many different people and the sum of all their views. The principle approaches to the study of public opinion may be divided into 4 categories:•quantitative measurement of opinion distributions;•investigation of the internal relationships among the individual opinions that make up public opinion on an issue;•description or analysis of the public role of public opinion;•study both of the communication media that disseminate the ideas on which opinions are based and of the uses that propagandists and other manipulators make of these media.Concepts of public opinion Public opinion as a concept gained credence with the rise of 'public' in the eighteenth century.
Search Engine	Search Engine was a weekly Canadian radio show that aired on CBC Radio One, then a dedicated podcast available on CBC.ca, and now a podcast on TVOntario's website, tvo.org. It is hosted by Jesse Brown, who also co-produces the show with Geoff Siskind and Andrew Parker. Cory Doctorow, novelist and editor of Boing Boing, is also a regular contributor.
Independent contractor	An independent contractor is a natural person, business, or corporation that provides goods or services to another entity under terms specified in a contract or within a verbal agreement. Unlike an employee, an independent contractor does not work regularly for an employer but works as and when required, during which time he or she may be subject to the Law of Agency. Independent contractors are usually paid on a freelance basis.
Credential	A credential is an attestation of qualification, competence, or authority issued to an individual by a third party with a relevant or de facto authority or assumed competence to do so. Examples of credentials include academic diplomas, academic degrees, certifications, security clearances, identification documents, badges, passwords, user names, keys, powers of attorney, and so on. Sometimes publications, such as scientific papers or books, may be viewed as similar to credentials by some people, especially if the publication was peer reviewed or made in a well-known journal or reputable publisher.
Intercultural	Cross-cultural communication (also frequently referred to as intercultural communication, which is also used in a different sense, though) is a field of study that looks at how people from differing cultural backgrounds communicate, in similar and different ways among themselves, and how they endeavour to communicate across cultures. Origins During the Cold War, the United States economy was largely self-contained because the world was polarized into two separate and competing powers: the east and west.

Chapter 9. Writing Negative Messages

Intercultural communication	Intercultural communication is a form of global communication. It is used to describe the wide range of communication problems that naturally appear within an organization made up of individuals from different religious, social, ethnic, and educational backgrounds. Intercultural communication is sometimes used synonymously with cross-cultural communication.
Buffer	A buffer in GIS is a zone around a map feature measured in units of distance or time. A buffer is useful for proximity analysis. A buffer is an area defined by the bounding region determined by a set of points at a specified maximum distance from all nodes along segments of an object.
Job interview	A job interview is a process in which a potential employee is evaluated by an employer for prospective employment in their company, organization, or firm. During this process, the employer hopes to determine whether or not the applicant is suitable for the role. A job interview typically precedes the hiring decision, and is used to evaluate the candidate.
Social network	A social network is a social structure made up of a set of actors (such as individuals or organizations) and the dyadic ties between these actors. The social network perspective provides a clear way of analyzing the structure of whole social entities. The study of these structures uses social network analysis to identify local and global patterns, locate influential entities, and examine network dynamics.
Podcasts	A podcast is a type of digital media consisting of an episodic series of audio radio, video, PDF, or ePub files subscribed to and downloaded through web syndication or streamed online to a computer or mobile device. The word is a neologism derived from 'broadcast' and 'pod' from the success of the iPod, as podcasts are often listened to on portable media players. In the context of Apple devices, the term 'Podcasts' refers to the audio and video version of podcasts, whereas the textual version of podcasts are classified under the app known as Newsstand.
Cultural identity	Cultural identity is the identity of a group or culture, or of an individual as far as one is influenced by one's belonging to a group or culture. Cultural identity is similar to and has overlaps with, but is not synonymous with, identity politics. Description Various modern cultural studies and social theories have investigated cultural identity.
Defamation	Defamation--also called calumny, vilification, traducement, slander (for transitory statements), and libel (for written, broadcast, or otherwise published words)--is the communication of a statement that makes a claim, expressly stated or implied to be factual, that may give an individual, business, product, group, government, or nation a negative image.

	This can be also any disparaging statement made by one person about another, which is communicated or published, whether true or false, depending on legal state. In Common Law it is usually a requirement that this claim be false and that the publication is communicated to someone other than the person defamed (the claimant).
Information technology	Information technology is a branch of engineering dealing with the use of computers and telecommunications equipment to store, retrieve, transmit and manipulate data. The term is commonly used as a synonym for computers and computer networks, but it also encompasses other information distribution technologies such as television and telephones. Some of the modern and emerging fields of Information technology are next generation web technologies, bioinformatics, cloud computing, global information systems, large-scale knowledge bases, etc.
Labor force	Normally, the labor force of a country consists of everyone of working age (typically above a certain age (around 14 to 16) and below retirement (around 65) who are participating workers, that is people actively employed or seeking employment. People not counted include students, retired people, stay-at-home parents, people in prisons or similar institutions, people employed in jobs or professions with unreported income, as well as discouraged workers who cannot find work. In the United States, the unemployment rate is estimated by a household survey called the Current Population Survey, conducted monthly by the Federal Bureau of Labor Statistics.
Crisis communication	Crisis communication is sometimes considered a sub-specialty of the public relations profession that is designed to protect and defend an individual, company, or organization facing a public challenge to its reputation. These challenges may come in the form of an investigation from a government agency, a criminal allegation, a media inquiry, a shareholders lawsuit, a violation of environmental regulations, or any of a number of other scenarios involving the legal, ethical, or financial standing of the entity. Crisis communication professionals preach that an organization's reputation is often its most valuable asset.
Crisis management	Crisis management is the process by which an organization deals with a major event that threatens to harm the organization, its stakeholders, or the general public. The study of crisis management originated with the large scale industrial and environmental disasters in the 1980s. Three elements are common to most definitions of crisis: (a) a threat to the organization, (b) the element of surprise, and (c) a short decision time.

Chapter 9. Writing Negative Messages

Financial Services	Financial services are the economic services provided by the finance industry, which encompasses a broad range of organizations that manage money, including credit unions, banks, credit card companies, insurance companies, consumer finance companies, stock brokerages, investment funds and some government sponsored enterprises. As of 2004, the financial services industry represented 20% of the market capitalization of the S&P 500 in the United States. History of financial services
	The term 'financial services' became more prevalent in the United States partly as a result of the Gramm-Leach-Bliley Act of the late 1990s, which enabled different types of companies operating in the U.S. financial services industry at that time to merge.
Proposal	A business proposal is a written offer from a seller to a prospective buyer. Business proposals are often a key step in the complex sales process--i.e., whenever a buyer considers more than price in a purchase.
	There are three distinct categories of business proposals:•formally solicited•informally solicited•unsolicited.
	Solicited proposals are written in response to published requirements, contained in a Request for Proposal (RFP), Request for Quotation (RFQ), Request for Information (RFI) or an Invitation For Bid (IFB).

1. An _____ is a natural person, business, or corporation that provides goods or services to another entity under terms specified in a contract or within a verbal agreement. Unlike an employee, an _____ does not work regularly for an employer but works as and when required, during which time he or she may be subject to the Law of Agency. _____s are usually paid on a freelance basis.

   a. Induction
   b. Independent contractor
   c. INGRADA
   d. Integrity Inventory

2. . A _____ is a social structure made up of a set of actors (such as individuals or organizations) and the dyadic ties between these actors. The _____ perspective provides a clear way of analyzing the structure of whole social entities. The study of these structures uses _____ analysis to identify local and global patterns, locate influential entities, and examine network dynamics.

   a. Jimmy Carter

   b. Capaware

   c. Cartographic generalization

   d. Social network

3. A _____ is an attestation of qualification, competence, or authority issued to an individual by a third party with a relevant or de facto authority or assumed competence to do so.

Examples of _____s include academic diplomas, academic degrees, certifications, security clearances, identification documents, badges, passwords, user names, keys, powers of attorney, and so on. Sometimes publications, such as scientific papers or books, may be viewed as similar to _____s by some people, especially if the publication was peer reviewed or made in a well-known journal or reputable publisher.

   a. Crossdisciplinarity

   b. Democratization of knowledge

   c. Descriptive knowledge

   d. Credential

4. A _____ is a process in which a potential employee is evaluated by an employer for prospective employment in their company, organization, or firm. During this process, the employer hopes to determine whether or not the applicant is suitable for the role.

A _____ typically precedes the hiring decision, and is used to evaluate the candidate.

   a. Jimmy Carter

   b. Capaware

   c. Cartographic generalization

   d. Job interview

5. . _____ is the aggregate of individual attitudes or beliefs held by the adult population. _____ can also be defined as the complex collection of opinions of many different people and the sum of all their views.

The principle approaches to the study of _____ may be divided into 4 categories:•quantitative measurement of opinion distributions;•investigation of the internal relationships among the individual opinions that make up _____ on an issue;•description or analysis of the public role of _____;•study both of the communication media that disseminate the ideas on which opinions are based and of the uses that propagandists and other manipulators make of these media.Concepts of _____

_____ as a concept gained credence with the rise of 'public' in the eighteenth century.

   a. Questionnaire

   b. Sampling

c. Public opinion

d. Self-report study

1. b

2. d

3. d

4. d

5. c

You can take the complete Chapter Practice Test

for Chapter 9. Writing Negative Messages
on all key terms, persons, places, and concepts.

Online 99 Cents

http://www.epub2.4.21191.9.cram101.com/

Use www.Cram101.com for all your study needs

including Cram101's online interactive problem solving labs in

chemistry, statistics, mathematics, and more.

CHAPTER OUTLINE: KEY TERMS, PEOPLE, PLACES, CONCEPTS

Crowdsourcing

Demographics

Psychographic

Public opinion

Social media

Social network

Statement

Affiliation

Information technology

Search Engine

Self-actualization

Cultural identity

Independent contractor

Marketing

Search engine marketing

Search engine optimization

Social networking

Action item

Resignation

Deductive reasoning

Hasty generalization

Inductive reasoning

Request for production

Financial Reporting

Objection

Production

Intercultural

Intercultural communication

Marketing Intelligence

Public speaking

Credential

Participation

Content

Content marketing

Connotation

CHAPTER HIGHLIGHTS & NOTES: KEY TERMS, PEOPLE, PLACES, CONCEPTS

Crowdsourcing	Crowdsourcing is a process that involves outsourcing tasks to a distributed group of people. This process can occur both online and offline. The difference between crowdsourcing and ordinary outsourcing is that a task or problem is outsourced to an undefined public rather than a specific body, such as paid employees.
Demographics	Demographics are current statistical characteristics of a population. These types of data are used widely in sociology (and especially in the subfield of demography), public policy, and marketing. Commonly examined demographics include gender, race, age, disabilities, mobility, home ownership, employment status, and even location.
Psychographic	In the fields of marketing, demographics, opinion research, and social research in general, psychographic variables are any attributes relating to personality, values, attitudes, interests, or lifestyles. They are also called IAO variables (for Interests, Activities, and Opinions). They can be contrasted with demographic variables (such as age and gender), behavioral variables (such as usage rate or loyalty), and firmographic variables (such as industry, seniority and functional area).
Public opinion	Public opinion is the aggregate of individual attitudes or beliefs held by the adult population. Public opinion can also be defined as the complex collection of opinions of many different people and the sum of all their views.
	The principle approaches to the study of public opinion may be divided into 4 categories:•quantitative measurement of opinion distributions;•investigation of the internal relationships among the individual opinions that make up public opinion on an issue;•description or analysis of the public role of public opinion;•study both of the communication media that disseminate the ideas on which opinions are based and of the uses that propagandists and other manipulators make of these media.Concepts of public opinion
	Public opinion as a concept gained credence with the rise of 'public' in the eighteenth century.
Social media	Social media includes web- and mobile-based technologies which are used to turn communication into interactive dialogue among organizations, communities, and individuals. Andreas Kaplan and Michael Haenlein define social media as 'a group of Internet-based applications that build on the ideological and technological foundations of Web 2.0, and that allow the creation and exchange of user-generated content.' When the technologies are in place, social media is ubiquitously accessible, and enabled by scalable communication techniques.
	Social media technologies take on many different forms including magazines, Internet forums, weblogs, social blogs, microblogging, wikis, social networks, podcasts, photographs or pictures, video, rating and social bookmarking.

Chapter 10. Writing Persuasive Messages

Social network	A social network is a social structure made up of a set of actors (such as individuals or organizations) and the dyadic ties between these actors. The social network perspective provides a clear way of analyzing the structure of whole social entities. The study of these structures uses social network analysis to identify local and global patterns, locate influential entities, and examine network dynamics.
Statement	In logic a statement is either (a) a meaningful declarative sentence that is either true or false, or (b) what is asserted or made by the use of a declarative sentence. In the latter case, a statement is distinct from a sentence in that a sentence is only one formulation of a statement, whereas there may be many other formulations expressing the same statement. Philosopher of language, Peter Strawson advocated the use of the term 'statement' in sense (b) in preference to proposition.
Affiliation	In law, affiliation is the term to describe a partnership between two or more parties. Affiliation procedures in England In England a number of statutes on the subject have been passed, the chief being the Bastardy Act of the Parliament of 1845, and the Bastardy Laws Amendment Acts of 1872 and 1873. The mother of a bastard may summon the putative father to petty sessions within 12 months of the birth (or at any later time if he is proved to have contributed to the child's support within 12 months after the birth), and the justices, as after hearing evidence on both sides, may, if the mother's evidence be corroborated in some material particular, adjudge the man to be the putative father] of the child, and order him to pay a sum not exceeding five shillings a week for its maintenance, together with a sum for expenses incidental to the birth, or the funeral expenses, if it has died before the date of order, and the costs of the proceedings. ceases to be valid after the child reaches the age of 13, but the justices (also referred to as Gold writers under these circumstances) may in the order direct the payments to be continued until the child is 16 years of age.
Information technology	Information technology is a branch of engineering dealing with the use of computers and telecommunications equipment to store, retrieve, transmit and manipulate data. The term is commonly used as a synonym for computers and computer networks, but it also encompasses other information distribution technologies such as television and telephones. Some of the modern and emerging fields of Information technology are next generation web technologies, bioinformatics, cloud computing, global information systems, large-scale knowledge bases, etc.
Search Engine	Search Engine was a weekly Canadian radio show that aired on CBC Radio One, then a dedicated podcast available on CBC.ca, and now a podcast on TVOntario's website, tvo.org.

It is hosted by Jesse Brown, who also co-produces the show with Geoff Siskind and Andrew Parker. Cory Doctorow, novelist and editor of Boing Boing, is also a regular contributor.

Self-actualization	Self-actualization is a term that has been used in various psychology theories, often in slightly different ways. The term was originally introduced by the organismic theorist Kurt Goldstein for the motive to realize one's full potential. In his view, it is the organism's master motive, the only real motive: 'the tendency to actualize itself as fully as possible is the basic drive...the drive of self-actualization.' Carl Rogers similarly wrote of 'the curative force in psychotherapy - man's tendency to actualize himself, to become his potentialities...to express and activate all the capacities of the organism.' However, the concept was brought most fully to prominence in Abraham Maslow's hierarchy of needs theory as the final level of psychological development that can be achieved when all basic and mental needs are fulfilled and the 'actualization' of the full personal potential takes place.
Cultural identity	Cultural identity is the identity of a group or culture, or of an individual as far as one is influenced by one's belonging to a group or culture. Cultural identity is similar to and has overlaps with, but is not synonymous with, identity politics. Description Various modern cultural studies and social theories have investigated cultural identity.
Independent contractor	An independent contractor is a natural person, business, or corporation that provides goods or services to another entity under terms specified in a contract or within a verbal agreement. Unlike an employee, an independent contractor does not work regularly for an employer but works as and when required, during which time he or she may be subject to the Law of Agency. Independent contractors are usually paid on a freelance basis.
Marketing	Marketing is 'the activity, set of institutions, and processes for creating, communicating, delivering, and exchanging offerings that have value for customers, clients, partners, and society at large.' For business to consumer marketing, it is 'the process by which companies create value for customers and build strong customer relationships, in order to capture value from customers in return'. For business to business marketing it is creating value, solutions, and relationships either short term or long term with a company or brand. It generates the strategy that underlies sales techniques, business communication, and business developments.
Search engine marketing	Search engine marketing is a form of internet marketing that involves the promotion of websites by increasing their visibility in search engine results pages (SERPs) through optimization (both on-page and off-page) as well as through advertising (paid placements, contextual advertising, and paid inclusions).

	Depending on the context, Search engine marketing can be an umbrella term for various means of marketing a website including search engine optimization (SEO), which adjusts or rewrites website content to achieve a higher ranking in search engine results pages, or it may contrast with PPC, focusing on only paid components.
	In 2008, North American advertisers spent US$13.5 billion on search engine marketing.
Search engine optimization	Search engine optimization is the process of improving the visibility of a website or a web page in a search engine's 'natural,' or un-paid ('organic' or 'algorithmic'), search results. In general, the earlier , and more frequently a site appears in the search results list, the more visitors it will receive from the search engine's users. Search engine optimization may target different kinds of search, including image search, local search, video search, academic search, news search and industry-specific vertical search engines.
Social networking	A social networking service is an online service, platform, or site that focuses on facilitating the building of social networks or social relations among people who, for example, share interests, activities, backgrounds, or real-life connections. A social network service consists of a representation of each user (often a profile), his/her social links, and a variety of additional services. Most social network services are web-based and provide means for users to interact over the Internet, such as e-mail and instant messaging.
Action item	In management, an action item is a documented event, task, activity, or action that needs to take place. Action items are discrete units that can be handled by a single person.
	Action items are usually created during a discussion by a group of people who are meeting about one or more topics and during the discussion it is discovered that some kind of action is needed.
Resignation	A resignation is the formal act of giving up or quitting one's office or position. It can also refer to the act of admitting defeat in a game like chess, indicated by the resigning player declaring 'I resign', turning his king on its side, extending his hand, or stopping the chess clock. A resignation can occur when a person holding a position gained by election or appointment steps down, but leaving a position upon the expiration of a term is not considered resignation.
Deductive reasoning	Deductive reasoning, is the process of reasoning from one or more general statements regarding what is known to reach a logically certain conclusion. Deductive reasoning involves using given true premises to reach a conclusion that is also true. Deductive reasoning contrasts with inductive reasoning in that a specific conclusion is arrived at from a general principle.

Hasty generalization	Hasty generalization is a logical fallacy of faulty generalization by reaching an inductive generalization based on insufficient evidence -- essentially making a hasty conclusion without considering all of the variables. In statistics, it may involve basing broad conclusions regarding the statistics of a survey from a small sample group that fails to sufficiently represent an entire population. Its opposite fallacy is called slothful induction, or denying the logical conclusion of an inductive argument (e.g. 'it was just a coincidence').
Inductive reasoning	Inductive reasoning, is a kind of reasoning that constructs or evaluates propositions that are abstractions of observations of individual instances of members of the same class. Inductive reasoning contrasts with deductive reasoning in that a general conclusion is arrived at by specific examples. The philosophical definition of inductive reasoning is much more nuanced than simple progression from particular/individual instances to wider generalizations.
Request for production	A request for production is a legal request for documents, electronically stored information, or other tangible items. In civil procedure, during the discovery phase of litigation, a party to a lawsuit may request that another party provide any documents that it has that pertain to the subject matter of the lawsuit. For example, a party in a court case may obtain copies of e-mail messages sent by employees of the opposing party.
Financial Reporting	Financial reporting is the process of preparing and distributing financial information to users of such information in various forms. The most common format of formal financial reporting are financial statements. Financial statements are prepared in accordance with rigorously applied standards defined by professional accounting bodies developed according to the legal and professional framework of a specific locale.
Objection	In the law of the United States of America, an objection is a formal protest raised in court during a trial to disallow a witness's testimony or other evidence which would be in violation of the rules of evidence or other procedural law. An objection is typically raised after the opposing party asks a question of the witness, but before the witness can answer, or when the opposing party is about to enter something into evidence. The judge then makes a ruling on whether the objection is 'sustained' (the judge agrees with the objection and disallows the question, testimony, or evidence) or 'overruled' (the judge disagrees with the objection and allows the question, testimony, or evidence).
Production	In economics, production is the act of creating output, a good or service which has value and contributes to the utility of individuals. The act may or may not include factors of production other than labor. Any effort directed toward the realization of a desired product or service is a 'productive' effort and the performance of such act is production.

Chapter 10. Writing Persuasive Messages

Intercultural	Cross-cultural communication (also frequently referred to as intercultural communication, which is also used in a different sense, though) is a field of study that looks at how people from differing cultural backgrounds communicate, in similar and different ways among themselves, and how they endeavour to communicate across cultures. Origins During the Cold War, the United States economy was largely self-contained because the world was polarized into two separate and competing powers: the east and west. However, changes and advancements in economic relationships, political systems, and technological options began to break down old cultural barriers.
Intercultural communication	Intercultural communication is a form of global communication. It is used to describe the wide range of communication problems that naturally appear within an organization made up of individuals from different religious, social, ethnic, and educational backgrounds. Intercultural communication is sometimes used synonymously with cross-cultural communication.
Marketing Intelligence	Marketing Intelligence is the information relevant to a company's markets, gathered and analyzed specifically for the purpose of accurate and confident decision-making in determining market opportunity, market penetration strategy, and market development metrics. Marketing intelligence is necessary when entering a foreign market. Marketing Intelligence is not the same as Market Intelligence (MARKINT).
Public speaking	Public speaking is the process of speaking to a group of people in a structured, deliberate manner intended to inform, influence, or entertain the listeners. It is closely allied to 'presenting', although the latter has more of a commercial advertisement. In public speaking, as in any form of communication, there are five basic elements, often expressed as 'who is saying what to whom using what medium with what effects?' The purpose of public speaking can range from simply transmitting information, to motivating people to act, to simply telling a story.
Credential	A credential is an attestation of qualification, competence, or authority issued to an individual by a third party with a relevant or de facto authority or assumed competence to do so. Examples of credentials include academic diplomas, academic degrees, certifications, security clearances, identification documents, badges, passwords, user names, keys, powers of attorney, and so on. Sometimes publications, such as scientific papers or books, may be viewed as similar to credentials by some people, especially if the publication was peer reviewed or made in a well-known journal or reputable publisher.
Participation	In finance, 'participation' is an ownership interest in a mortgage or other loan.

In particular, loan participation is a cooperation of multiple lenders to issue a loan (known as participation loan) to one borrower. This is usually done in order to reduce individual risks of the lenders.

Content

In mathematics, a content is a real function μ defined on a field of sets $\mathcal{A}$ such that

$$\mu(A) \in [0, \infty] \text{ whenever } A \in \mathcal{A}. \quad \mu(\varnothing) = 0.$$

$$\mu(A_1 \cup A_2) = \mu(A_1) + \mu(A_2) \text{ whenever } A_1, A_2 \in \mathcal{A} \text{ and } A_1 \cap A_2 = \varnothing.$$

A very important type of content is a measure, which is a σ-additive content defined on a σ-field. Every measure is a content, but not vice-versa.

Content marketing

Content marketing is an umbrella term encompassing all marketing formats that involve the creation and sharing of content in order to engage current and potential consumer bases. Content marketing subscribes to the notion that delivering high-quality, relevant and valuable information to prospects and customers drives profitable consumer action. Content marketing has benefits in terms of retaining reader attention and improving brand loyalty.

Connotation

A connotation is a commonly understood subjective cultural or emotional association that some word or phrase carries, in addition to the word's or phrase's explicit or literal meaning, which is its denotation.

A connotation is frequently described as either positive or negative, with regards to its pleasing or displeasing emotional connection. For example, a stubborn person may be described as being either strong-willed or pig-headed; although these have the same literal meaning , strong-willed connotes admiration for the level of someone's will (a positive connotation), while pig-headed connotes frustration in dealing with someone (a negative connotation).

Chapter 10. Writing Persuasive Messages

1. An _____ is a natural person, business, or corporation that provides goods or services to another entity under terms specified in a contract or within a verbal agreement. Unlike an employee, an _____ does not work regularly for an employer but works as and when required, during which time he or she may be subject to the Law of Agency. _____s are usually paid on a freelance basis.

   a. Induction
   b. Independent contractor
   c. INGRADA
   d. Integrity Inventory

2. In economics, _____ is the act of creating output, a good or service which has value and contributes to the utility of individuals. The act may or may not include factors of _____ other than labor. Any effort directed toward the realization of a desired product or service is a 'productive' effort and the performance of such act is _____.

   a. Production
   b. Term
   c. Omnibus hearing
   d. Open verdict

3. _____ is the information relevant to a company's markets, gathered and analyzed specifically for the purpose of accurate and confident decision-making in determining market opportunity, market penetration strategy, and market development metrics. _____ is necessary when entering a foreign market. _____ is not the same as Market Intelligence (MARKINT).

   a. Marketing Intelligence
   b. Marketing mix
   c. Marketing mix for product software
   d. Marketing myopia

4. _____, is the process of reasoning from one or more general statements regarding what is known to reach a logically certain conclusion. _____ involves using given true premises to reach a conclusion that is also true. _____ contrasts with inductive reasoning in that a specific conclusion is arrived at from a general principle.

   a. Lateral thinking
   b. Jimmy Carter
   c. Multifactor
   d. Deductive reasoning

5. . In the law of the United States of America, an _____ is a formal protest raised in court during a trial to disallow a witness's testimony or other evidence which would be in violation of the rules of evidence or other procedural law. An _____ is typically raised after the opposing party asks a question of the witness, but before the witness can answer, or when the opposing party is about to enter something into evidence.

The judge then makes a ruling on whether the _____ is 'sustained' (the judge agrees with the _____ and disallows the question, testimony, or evidence) or 'overruled' (the judge disagrees with the _____ and allows the question, testimony, or evidence).

a. Offer of proof
b. Objection
c. Omnibus hearing
d. Open verdict

1. b
2. a
3. a
4. d
5. b

You can take the complete Chapter Practice Test

for Chapter 10. Writing Persuasive Messages
on all key terms, persons, places, and concepts.

Online 99 Cents

http://www.epub2.4.21191.10.cram101.com/

Use www.Cram101.com for all your study needs

including Cram101's online interactive problem solving labs in

chemistry, statistics, mathematics, and more.

Research participant

Statement

Digital rights

Gap analysis

Intellectual property

Misrepresentation

Credential

Primary research

Reliability

Secondary research

Knowledge management

Management system

Independent contractor

Electronic business

Newspaper

Proposal

Search Engine

Cultural identity

Social tagging

_____ | Law firm

_____ | Enterprise search

_____ | Job interview

_____ | Social bookmarking

_____ | Request for production

_____ | Fair use

_____ | Leading question

_____ | Sales pitch

_____ | Social media

_____ | Focus group

_____ | Median

_____ | Financial Reporting

_____ | Trend analysis

Research participant	A research participant, trial, or study participant or subject, is a person who participates in human subject research by being the target of observation by researchers.
	Researchers who conduct human subject research should afford special rights to research participants. Research participants should expect the following:•to be the target of beneficence•to experience research justice•to get respect for persons•to have privacy for research participants.
Statement	In logic a statement is either (a) a meaningful declarative sentence that is either true or false, or (b) what is asserted or made by the use of a declarative sentence. In the latter case, a statement is distinct from a sentence in that a sentence is only one formulation of a statement, whereas there may be many other formulations expressing the same statement.
	Philosopher of language, Peter Strawson advocated the use of the term 'statement' in sense (b) in preference to proposition.
Digital rights	The term digital rights describes the protections that allow individuals to access, use, create, and publish digital media or to access and use computers, other electronic devices, or communications networks. The term is particularly related to the protection and realization of existing rights, such as the right to privacy or freedom of expression, in the context of new digital technologies, especially the Internet.
	In several countries, including Estonia, France, Finland, Greece and Spain, Internet access is considered a human right.
Gap analysis	In business and economics, gap analysis is a tool that helps companies compare actual performance with potential performance. At its core are two questions: 'Where are we?' and 'Where do we want to be?' If a company or organization does not make the best use of current resources, or foregoes investment in capital or technology, it may produce or perform below its potential. This concept is similar to the base case of being below the production possibilities frontier.
Intellectual property	Intellectual property is a term referring to a number of distinct types of expressions for which a set of rights are recognized under the corresponding fields of law. Under intellectual property law, owners are granted certain exclusive rights to various markets, machines, musical, literary, and artistic works; discoveries and inventions; and applications. Common types of intellectual property rights include copyrights, trademarks, patents, industrial design rights, and trade secrets in some jurisdictions.
Misrepresentation	Misrepresentation is a contract law concept. It means a false statement of fact made by one party to another party, which has the effect of inducing that party into the contract.

Chapter 11. Finding, Evaluating, and Processing Information

Credential	A credential is an attestation of qualification, competence, or authority issued to an individual by a third party with a relevant or de facto authority or assumed competence to do so.
	Examples of credentials include academic diplomas, academic degrees, certifications, security clearances, identification documents, badges, passwords, user names, keys, powers of attorney, and so on. Sometimes publications, such as scientific papers or books, may be viewed as similar to credentials by some people, especially if the publication was peer reviewed or made in a well-known journal or reputable publisher.
Primary research	Primary research consists of the collection of original primary data. It is often undertaken after the researcher has gained some insight into the issue by reviewing secondary research or by analyzing previously collected primary data. It can be accomplished through various methods, including questionnaires and telephone interviews in market research, or experiments and direct observations in the physical sciences, amongst others.
Reliability	Reliability of semiconductor devices can be summarized as follows:•Semiconductor devices are very sensitive to impurities and particles. Therefore, to manufacture these devices it is necessary to manage many processes while accurately controlling the level of impurities and particles. The finished product quality depends upon the many layered relationship of each interacting substance in the semiconductor, including metallization, chip material (list of semiconductor materials) and package.•The problems of micro-processes, and thin films and must be fully understood as they apply to metallization and bonding wire bonding.
Secondary research	Secondary research involves the summary, collation and/or synthesis of existing research rather than primary research, where data is collected from, for example, research subjects or experiments.
	The term is widely used in medical research and in market research. The principal methodology in medical secondary research is the systematic review, commonly using meta-analytic statistical techniques, although other methods of synthesis, like realist reviews and meta-narrative reviews, have been developed in recent years.
Knowledge management	Knowledge management comprises a range of strategies and practices used in an organization to identify, create, represent, distribute, and enable adoption of insights and experiences. Such insights and experiences comprise knowledge, either embodied in individuals or embedded in organizations as processes or practices.
	An established discipline since 1991 , KM includes courses taught in the fields of business administration, information systems, management, and library and information sciences (Alavi & Leidner 1999).

Management system	A management system is the framework of processes and procedures used to ensure that an organization can fulfill all tasks required to achieve its objectives. For instance, an environmental management system enables organizations to improve their environmental performance through a process of continuous improvement. An oversimplification is 'Plan, Do, Check, Act'.
Independent contractor	An independent contractor is a natural person, business, or corporation that provides goods or services to another entity under terms specified in a contract or within a verbal agreement. Unlike an employee, an independent contractor does not work regularly for an employer but works as and when required, during which time he or she may be subject to the Law of Agency. Independent contractors are usually paid on a freelance basis.
Electronic business	Electronic business, commonly referred to as 'eBusiness' or 'e-business', or an internet business, may be defined as the application of information and communication technologies (ICT) in support of all the activities of business. Commerce constitutes the exchange of products and services between businesses, groups and individuals and can be seen as one of the essential activities of any business. Electronic commerce focuses on the use of ICT to enable the external activities and relationships of the business with individuals, groups and other businesses.
Newspaper	A newspaper is a scheduled publication containing news of current events, informative articles, diverse features, editorials, and advertising. It usually is printed on relatively inexpensive, low-grade paper such as newsprint. By 2007, there were 6580 daily newspapers in the world selling 395 million copies a day.
Proposal	A business proposal is a written offer from a seller to a prospective buyer. Business proposals are often a key step in the complex sales process--i.e., whenever a buyer considers more than price in a purchase. There are three distinct categories of business proposals:•formally solicited•informally solicited•unsolicited. Solicited proposals are written in response to published requirements, contained in a Request for Proposal (RFP), Request for Quotation (RFQ), Request for Information (RFI) or an Invitation For Bid (IFB).
Search Engine	Search Engine was a weekly Canadian radio show that aired on CBC Radio One, then a dedicated podcast available on CBC.ca, and now a podcast on TVOntario's website, tvo.org. It is hosted by Jesse Brown, who also co-produces the show with Geoff Siskind and Andrew Parker.

Chapter 11. Finding, Evaluating, and Processing Information

Cultural identity	Cultural identity is the identity of a group or culture, or of an individual as far as one is influenced by one's belonging to a group or culture. Cultural identity is similar to and has overlaps with, but is not synonymous with, identity politics. Description Various modern cultural studies and social theories have investigated cultural identity.
Social tagging	A folksonomy is a system of classification derived from the practice and method of collaboratively creating and managing tags to annotate and categorize content; this practice is also known as collaborative tagging, social classification, social indexing, and social tagging. Folksonomy, a term coined by Thomas Vander Wal, is a portmanteau of folk and taxonomy. Folksonomies became popular on the Web around 2004 as part of social software applications such as social bookmarking and photograph annotation.
Law firm	A law firm is a business entity formed by one or more lawyers to engage in the practice of law. The primary service rendered by a law firm is to advise clients (individuals or corporations) about their legal rights and responsibilities, and to represent clients in civil or criminal cases, business transactions, and other matters in which legal advice and other assistance are sought. Law firms are organized in a variety of ways, depending on the jurisdiction in which the firm practices.
Enterprise search	Enterprise search is the practice of making content from multiple enterprise-type sources, such as databases and intranets, searchable to a defined audience. 'Enterprise Search' is used to describe the software of search information within an enterprise (though the search function and its results may still be public). Enterprise search can be contrasted with web search, which applies search technology to documents on the open web, and desktop search, which applies search technology to the content on a single computer.
Job interview	A job interview is a process in which a potential employee is evaluated by an employer for prospective employment in their company, organization, or firm. During this process, the employer hopes to determine whether or not the applicant is suitable for the role. A job interview typically precedes the hiring decision, and is used to evaluate the candidate.
Social bookmarking	Social bookmarking is a method for Internet users to organize, store, manage and search for bookmarks of resources online. Many online bookmark management services have launched since 1996; Delicious, founded in 2003, popularized the terms 'social bookmarking' and 'tagging'.

Request for production	A request for production is a legal request for documents, electronically stored information, or other tangible items. In civil procedure, during the discovery phase of litigation, a party to a lawsuit may request that another party provide any documents that it has that pertain to the subject matter of the lawsuit. For example, a party in a court case may obtain copies of e-mail messages sent by employees of the opposing party.
Fair use	Fair use is a limitation and exception to the exclusive right granted by copyright law to the author of a creative work. In United States copyright law, fair use is a doctrine that permits limited use of copyrighted material without acquiring permission from the rights holders. Examples of fair use include commentary, criticism, news reporting, research, teaching, library archiving and scholarship.
Leading question	In common law systems that rely on testimony by witnesses, a leading question is looking for. For example, this question is leading:•You were at KC's bar on the night of July 15, weren't you? It suggests that the witness was at KC's bar on the night in question. The same question in a non-leading form would be:•Where were you on the night of July 15? This form of question does not suggest to the witness the answer the examiner hopes to elicit.
Sales pitch	In selling technique, a sales pitch is a line of talk that attempts to persuade someone or something, with a planned sales presentation strategy of a product or service designed to initiate and close a sale of the product or service. A sales pitch is a planned presentation of a product or service designed to initiate and close a sale of the same product or service. A sales pitch is essentially designed to be either an introduction of a product or service to an audience who knows nothing about it, or a descriptive expansion of a product or service that an audience has already expressed interest in.
Social media	Social media includes web- and mobile-based technologies which are used to turn communication into interactive dialogue among organizations, communities, and individuals. Andreas Kaplan and Michael Haenlein define social media as 'a group of Internet-based applications that build on the ideological and technological foundations of Web 2.0, and that allow the creation and exchange of user-generated content.' When the technologies are in place, social media is ubiquitously accessible, and enabled by scalable communication techniques. Social media technologies take on many different forms including magazines, Internet forums, weblogs, social blogs, microblogging, wikis, social networks, podcasts, photographs or pictures, video, rating and social bookmarking.

Chapter 11. Finding, Evaluating, and Processing Information

Focus group	A focus group is a form of qualitative research in which a group of people are asked about their perceptions, opinions, beliefs, and attitudes towards a product, service, concept, advertisement, idea, or packaging. Questions are asked in an interactive group setting where participants are free to talk with other group members. The first focus groups were created at the Bureau of Applied Social Research in the USA, by associate director, sociologist Robert K. Merton.
Median	In statistics and probability theory, median is described as the numerical value separating the higher half of a sample, a population, or a probability distribution, from the lower half. The median of a finite list of numbers can be found by arranging all the observations from lowest value to highest value and picking the middle one. If there is an even number of observations, then there is no single middle value; the median is then usually defined to be the mean of the two middle values.
Financial Reporting	Financial reporting is the process of preparing and distributing financial information to users of such information in various forms. The most common format of formal financial reporting are financial statements. Financial statements are prepared in accordance with rigorously applied standards defined by professional accounting bodies developed according to the legal and professional framework of a specific locale.
Trend analysis	Trend Analysis is the practice of collecting information and attempting to spot a pattern, or trend, in the information. In some fields of study, the term 'trend analysis' has more formally defined meanings. Although trend analysis is often used to predict future events, it could be used to estimate uncertain events in the past, such as how many ancient kings probably ruled between two dates, based on data such as the average years which other known kings reigned.

1. . _____ involves the summary, collation and/or synthesis of existing research rather than primary research, where data is collected from, for example, research subjects or experiments.

    The term is widely used in medical research and in market research. The principal methodology in medical _____ is the systematic review, commonly using meta-analytic statistical techniques, although other methods of synthesis, like realist reviews and meta-narrative reviews, have been developed in recent years.

    a. Segmenting and positioning
    b. Seminar marketing
    c. Secondary research

2. A _____ is an attestation of qualification, competence, or authority issued to an individual by a third party with a relevant or de facto authority or assumed competence to do so.

   Examples of _____s include academic diplomas, academic degrees, certifications, security clearances, identification documents, badges, passwords, user names, keys, powers of attorney, and so on. Sometimes publications, such as scientific papers or books, may be viewed as similar to _____s by some people, especially if the publication was peer reviewed or made in a well-known journal or reputable publisher.

   a. Crossdisciplinarity
   b. Democratization of knowledge
   c. Credential
   d. Dispersed knowledge

3. _____ is a contract law concept. It means a false statement of fact made by one party to another party, which has the effect of inducing that party into the contract. For example, under certain circumstances, false statements or promises made by a seller of goods regarding the quality or nature of the product that the seller has may constitute _____.

   a. Jimmy Carter
   b. Misrepresentation
   c. Internet forum
   d. Multifactor

4. _____ is a term referring to a number of distinct types of expressions for which a set of rights are recognized under the corresponding fields of law. Under _____ law, owners are granted certain exclusive rights to various markets, machines, musical, literary, and artistic works; discoveries and inventions; and applications. Common types of _____ rights include copyrights, trademarks, patents, industrial design rights, and trade secrets in some jurisdictions.

   a. organizational learning
   b. Information economy
   c. Intellectual property
   d. Edith of Wilton

5. _____ was a weekly Canadian radio show that aired on CBC Radio One, then a dedicated podcast available on CBC.ca, and now a podcast on TVOntario's website, tvo.org. It is hosted by Jesse Brown, who also co-produces the show with Geoff Siskind and Andrew Parker. Cory Doctorow, novelist and editor of Boing Boing, is also a regular contributor.

   a. Security Now
   b. Search Engine
   c. This Week in Tech
   d. Tips from the Top Floor

1. c
2. c
3. b
4. c
5. b

You can take the complete Chapter Practice Test

for Chapter 11. Finding, Evaluating, and Processing Information
on all key terms, persons, places, and concepts.

Online 99 Cents

http://www.epub2.4.21191.11.cram101.com/

Use **www.Cram101.com** for all your study needs

including Cram101's online interactive problem solving labs in

chemistry, statistics, mathematics, and more.

Chapter 12. Designing Visual Communication

Virtual team

Visual communication

Social networking

Chartjunk

Element

Distortion

Orientation

Bar chart

Flowchart

Job interview

Line chart

Pie chart

Cultural identity

Linear regression

Regression analysis

Gantt chart

Pictogram

Social media

Tag cloud

	Organizational chart
	Computer-aided design
	Information graphics
	Information systems
	Statement
	Online shopping
	Sidebar

CHAPTER HIGHLIGHTS & NOTES: KEY TERMS, PEOPLE, PLACES, CONCEPTS

Virtual team	A virtual team is a group of individuals who work across time, space and organizational boundaries with links strengthened by webs of communication technology' Ale Ebrahim, N., Ahmed, S. & Taha, Z. in their recent (2009) literature review paper, added two key issues to definition of a virtual team 'as small temporary groups of geographically, organizationally and/ or time dispersed knowledge workers who coordinate their work predominantly with electronic information and communication technologies in order to accomplish one or more organization tasks' Members of virtual teams communicate electronically and may never meet face-to-face. Virtual teams are made possible by a proliferation of fiber optic technology that has significantly increased the scope of off-site communication.
Visual communication	Visual communication as the name suggests is communication through visual aid and is described as the conveyance of ideas and information in forms that can be read or looked upon. Visual communication in part or whole relies on vision, and is primarily presented or expressed with two dimensional images, it includes: signs, typography, drawing, graphic design, illustration, colour and electronic resources. It also explores the idea that a visual message accompanying text has a greater power to inform, educate, or persuade a person or audience.

Chapter 12. Designing Visual Communication

Social networking	A social networking service is an online service, platform, or site that focuses on facilitating the building of social networks or social relations among people who, for example, share interests, activities, backgrounds, or real-life connections. A social network service consists of a representation of each user (often a profile), his/her social links, and a variety of additional services. Most social network services are web-based and provide means for users to interact over the Internet, such as e-mail and instant messaging.
Chartjunk	Chartjunk refers to all visual elements in charts and graphs that are not necessary to comprehend the information represented on the graph, or that distract the viewer from this information. Markings and visual elements can be called chartjunk if they are not part of the minimum set of visuals necessary to communicate the information understandably. Examples of unnecessary elements which might be called chartjunk include heavy or dark grid lines, unnecessary text or inappropriately complex fontfaces, ornamented chart axes and display frames, pictures or icons within data graphs, ornamental shading and unnecessary dimensions.
Element	Under United States law, an element of a crime is one of a set of facts that must all be proven to convict a defendant of a crime. Before a court finds a defendant guilty of a criminal offense, the prosecution must present evidence that, even when opposed by any evidence the defense may choose to present, is credible and sufficient to prove beyond a reasonable doubt that the defendant committed each element of the particular crime charged. The component parts that make up any particular crime vary depending on the crime.
Distortion	A distortion is departure from the allocation of economic resources from the state in which each agent maximizes her own welfare.. A proportional wage-income tax, for instance, is distortionary, whereas a Lump-sum tax is not. In a competitive equilibrium, a proportional wage income tax discourages work.
Orientation	In mathematics, orientation is a geometric notion that in two dimensions allows one to say when a cycle goes around clockwise or counterclockwise, and in three dimensions when a figure is left-handed or right-handed. In linear algebra, the notion of orientation makes sense in arbitrary dimensions. In this setting, the orientation of an ordered basis is a kind of asymmetry that makes a reflection impossible to replicate by means of a simple rotation.
Bar chart	A bar chart is a chart with rectangular bars with lengths proportional to the values that they represent. The bars can be plotted vertically or horizontally. A vertical bar chart is sometimes called a column bar chart.
Flowchart	A flowchart is a type of diagram that represents an algorithm or process, showing the steps as boxes of various kinds, and their order by connecting these with arrows.

This diagrammatic representation can give a step-by-step solution to a given problem. Process operations are represented in these boxes, and arrows connecting them represent flow of control.

Job interview

A job interview is a process in which a potential employee is evaluated by an employer for prospective employment in their company, organization, or firm. During this process, the employer hopes to determine whether or not the applicant is suitable for the role.

A job interview typically precedes the hiring decision, and is used to evaluate the candidate.

Line chart

A line chart is a type of chart which displays information as a series of data points connected by straight line segments. It is a basic type of chart common in many fields. It is an extension of a scatter graph, and is created by connecting a series of points that represent individual measurements with line segments.

Pie chart

A pie chart is a circular chart divided into sectors, illustrating proportion. In a pie chart, the arc length of each sector (and consequently its central angle and area), is proportional to the quantity it represents. When angles are measured with 1 turn as unit then a number of percent is identified with the same number of centiturns.

Cultural identity

Cultural identity is the identity of a group or culture, or of an individual as far as one is influenced by one's belonging to a group or culture. Cultural identity is similar to and has overlaps with, but is not synonymous with, identity politics. Description

Various modern cultural studies and social theories have investigated cultural identity.

Linear regression

In statistics, linear regression is an approach to modelling the relationship between a scalar dependent variable y and one or more explanatory variables denoted X. The case of one explanatory variable is called simple regression. More than one explanatory variable is multiple regression. (This in turn should be distinguished from multivariate linear regression, where multiple correlated dependent variables are predicted, rather than a single scalar variable).

Regression analysis

In statistics, regression analysis includes many techniques for modeling and analyzing several variables, when the focus is on the relationship between a dependent variable and one or more independent variables. More specifically, regression analysis helps one understand how the typical value of the dependent variable changes when any one of the independent variables is varied, while the other independent variables are held fixed. Most commonly, regression analysis estimates the conditional expectation of the dependent variable given the independent variables -- that is, the average value of the dependent variable when the independent variables are held fixed.

Chapter 12. Designing Visual Communication

Gantt chart	A Gantt chart is a type of bar chart, developed by Henry Gantt, that illustrates a project schedule. Gantt charts illustrate the start and finish dates of the terminal elements and summary elements of a project. Terminal elements and summary elements comprise the work breakdown structure of the project.
Pictogram	A pictogram, is an ideogram that conveys its meaning through its pictorial resemblance to a physical object. Pictographs are often used in writing and graphic systems in which the characters are to considerable extent pictorial in appearance. Pictography is a form of writing which uses representational, pictorial drawings.
Social media	Social media includes web- and mobile-based technologies which are used to turn communication into interactive dialogue among organizations, communities, and individuals. Andreas Kaplan and Michael Haenlein define social media as 'a group of Internet-based applications that build on the ideological and technological foundations of Web 2.0, and that allow the creation and exchange of user-generated content.' When the technologies are in place, social media is ubiquitously accessible, and enabled by scalable communication techniques. Social media technologies take on many different forms including magazines, Internet forums, weblogs, social blogs, microblogging, wikis, social networks, podcasts, photographs or pictures, video, rating and social bookmarking.
Tag cloud	A tag cloud is a visual representation for text data, typically used to depict keyword metadata (tags) on websites, or to visualize free form text. Tags are usually single words, and the importance of each tag is shown with font size or color. This format is useful for quickly perceiving the most prominent terms and for locating a term alphabetically to determine its relative prominence.
Organizational chart	An organizational chart (often called organization chart, org chart, organigram(me), or organogram (me)) is a diagram that shows the structure of an organization and the relationships and relative ranks of its parts and positions/jobs. The term is also used for similar diagrams, for example ones showing the different elements of a field of knowledge or a group of languages. The French Encyclopédie published in France between 1751 and 1772 had one of the first organizational charts of knowledge in general.
Computer-aided design	Computer-aided design also known as computer-aided drafting or computer-aided design and drafting (CADD), is the use of computer systems to assist in the creation, modification, analysis, or optimization of a design. Computer-aided drafting describes the process of creating a technical drawing with the use of computer software.

Information graphics	Information graphics are graphic visual representations of information, data or knowledge. These graphics present complex information quickly and clearly, such as in signs, maps, journalism, technical writing, and education. With an information graphic, computer scientists, mathematicians, and statisticians develop and communicate concepts using a single symbol to process information.
Information systems	Information systems is the study of complementary networks of hardware and software that people and organizations use to collect, filter, process, create, and distribute data. The study bridges business and computer science using the theoretical foundations of information and computation to study various business models and related algorithmic processes within a computer science discipline. Computer Information System(s) (CIS) is a field studying computers and algorithmic processes, including their principles, their software and hardware designs, their applications, and their impact on society while IS emphasizes functionality over design.
Statement	In logic a statement is either (a) a meaningful declarative sentence that is either true or false, or (b) what is asserted or made by the use of a declarative sentence. In the latter case, a statement is distinct from a sentence in that a sentence is only one formulation of a statement, whereas there may be many other formulations expressing the same statement.

Philosopher of language, Peter Strawson advocated the use of the term 'statement' in sense (b) in preference to proposition. |
| Online shopping | Online shopping is a form of electronic commerce whereby consumers directly buy goods or services from a seller over the Internet without an intermediary service. An online shop, eshop, e-store, Internet shop, webshop, webstore, online store, or virtual store evokes the physical analogy of buying products or services at a bricks-and-mortar retailer or shopping centre. The process is called business-to-consumer (B2C) online shopping. |
| Sidebar | The sidebar is an area in a courtroom near the judge's bench where lawyers may be called to speak with the judge so that the jury cannot hear the conversation and/or they may speak off the record. Lawyers make a formal request by stating 'may I approach the bench?' or, simply 'may I approach?' to initiate a sidebar conference. If it is granted, then opposing counsel must be allowed to come forward and participate in the conversation. |

Chapter 12. Designing Visual Communication

1. In statistics, _____ includes many techniques for modeling and analyzing several variables, when the focus is on the relationship between a dependent variable and one or more independent variables. More specifically, _____ helps one understand how the typical value of the dependent variable changes when any one of the independent variables is varied, while the other independent variables are held fixed. Most commonly, _____ estimates the conditional expectation of the dependent variable given the independent variables -- that is, the average value of the dependent variable when the independent variables are held fixed.

    a. Seemingly unrelated regressions
    b. Spatial econometrics
    c. Regression analysis
    d. Survivorship bias

2. A _____ is a process in which a potential employee is evaluated by an employer for prospective employment in their company, organization, or firm. During this process, the employer hopes to determine whether or not the applicant is suitable for the role.

    A _____ typically precedes the hiring decision, and is used to evaluate the candidate.

    a. Job interview
    b. p-chart
    c. Multifactor
    d. Orthogonal Procrustes problem

3. In logic a _____ is either (a) a meaningful declarative sentence that is either true or false, or (b) what is asserted or made by the use of a declarative sentence. In the latter case, a _____ is distinct from a sentence in that a sentence is only one formulation of a _____, whereas there may be many other formulations expressing the same _____.

    Philosopher of language, Peter Strawson advocated the use of the term '_____' in sense (b) in preference to proposition.

    a. Statement
    b. Term logic
    c. Testability
    d. Tetralemma

4. . In statistics, _____ is an approach to modelling the relationship between a scalar dependent variable y and one or more explanatory variables denoted X. The case of one explanatory variable is called simple regression. More than one explanatory variable is multiple regression. (This in turn should be distinguished from multivariate _____, where multiple correlated dependent variables are predicted, rather than a single scalar variable).

    a. Local independence
    b. Linear regression
    c. Market facilitation index

5. A _____ is a type of bar chart, developed by Henry Gantt, that illustrates a project schedule. _____s illustrate the start and finish dates of the terminal elements and summary elements of a project. Terminal elements and summary elements comprise the work breakdown structure of the project.

a. Grandfather principle
b. Group information management
c. Master plan
d. Gantt chart

1. c
2. a
3. a
4. b
5. d

You can take the complete Chapter Practice Test

for Chapter 12. Designing Visual Communication
on all key terms, persons, places, and concepts.

Online 99 Cents

http://www.epub2.4.21191.12.cram101.com/

Use www.Cram101.com for all your study needs

including Cram101's online interactive problem solving labs in

chemistry, statistics, mathematics, and more.

CHAPTER OUTLINE: KEY TERMS, PEOPLE, PLACES, CONCEPTS

Job interview

Proposal

Social networking

Public opinion

Statement

Dashboard

Independent contractor

Receiver

Orientation

Cultural identity

P3M3

White paper

Business plan

Categories

Exit strategy

Marketing Intelligence

Marketing strategy

Mission statement

Content

	Information architecture
	Social media
	Virtual team
	Administration
	Small business
	Chunking
	Discovery
	Due diligence
	Failure analysis
	Market analysis
	Request for proposal
	Sales pitch
	Credential
	Annual report

Job interview	A job interview is a process in which a potential employee is evaluated by an employer for prospective employment in their company, organization, or firm. During this process, the employer hopes to determine whether or not the applicant is suitable for the role.
	A job interview typically precedes the hiring decision, and is used to evaluate the candidate.
Proposal	A business proposal is a written offer from a seller to a prospective buyer. Business proposals are often a key step in the complex sales process--i.e., whenever a buyer considers more than price in a purchase.
	There are three distinct categories of business proposals:•formally solicited•informally solicited•unsolicited.
	Solicited proposals are written in response to published requirements, contained in a Request for Proposal (RFP), Request for Quotation (RFQ), Request for Information (RFI) or an Invitation For Bid (IFB).
Social networking	A social networking service is an online service, platform, or site that focuses on facilitating the building of social networks or social relations among people who, for example, share interests, activities, backgrounds, or real-life connections. A social network service consists of a representation of each user (often a profile), his/her social links, and a variety of additional services. Most social network services are web-based and provide means for users to interact over the Internet, such as e-mail and instant messaging.
Public opinion	Public opinion is the aggregate of individual attitudes or beliefs held by the adult population. Public opinion can also be defined as the complex collection of opinions of many different people and the sum of all their views.
	The principle approaches to the study of public opinion may be divided into 4 categories:•quantitative measurement of opinion distributions;•investigation of the internal relationships among the individual opinions that make up public opinion on an issue;•description or analysis of the public role of public opinion;•study both of the communication media that disseminate the ideas on which opinions are based and of the uses that propagandists and other manipulators make of these media.Concepts of public opinion
	Public opinion as a concept gained credence with the rise of 'public' in the eighteenth century.
Statement	In logic a statement is either (a) a meaningful declarative sentence that is either true or false, or (b) what is asserted or made by the use of a declarative sentence. In the latter case, a statement is distinct from a sentence in that a sentence is only one formulation of a statement, whereas there may be many other formulations expressing the same statement.

Chapter 13. Planning Reports and Proposals

Dashboard	In management information systems, a dashboard is' 'An easy to read, often single page, real-time user interface, showing a graphical presentation of the current status (snapshot) and historical trends of an organization's Key Performance Indicators (KPIs) to enable instantaneous and informed decisions to be made at a glance.' ' For example, a manufacturing dashboard may show KPIs related to productivity such as number of parts manufactured, or number of failed quality inspections per hour. Similarly, a human resources dashboard may show KPIs related to staff recruitment, retention and composition, for example number of open positions, or average days or cost per recruitment. Dashboards should not be confused with scorecards.
Independent contractor	An independent contractor is a natural person, business, or corporation that provides goods or services to another entity under terms specified in a contract or within a verbal agreement. Unlike an employee, an independent contractor does not work regularly for an employer but works as and when required, during which time he or she may be subject to the Law of Agency. Independent contractors are usually paid on a freelance basis.
Receiver	In modulated ultrasound terminology, a receiver is a device that receives a modulated ultrasound signal and decodes it for use as sound, navigational-position information, etc. Its function is somewhat like that of a radio receiver
Orientation	In mathematics, orientation is a geometric notion that in two dimensions allows one to say when a cycle goes around clockwise or counterclockwise, and in three dimensions when a figure is left-handed or right-handed. In linear algebra, the notion of orientation makes sense in arbitrary dimensions. In this setting, the orientation of an ordered basis is a kind of asymmetry that makes a reflection impossible to replicate by means of a simple rotation.
Cultural identity	Cultural identity is the identity of a group or culture, or of an individual as far as one is influenced by one's belonging to a group or culture. Cultural identity is similar to and has overlaps with, but is not synonymous with, identity politics. Description Various modern cultural studies and social theories have investigated cultural identity.
P3M3	P3M3, Programme and Project Management Maturity Model is a reference guide for structured best practice. It breaks down the broad disciplines of portfolio, programme and project management into a hierarchy of Key Process Areas (KPAs). The hierarchical approach enables organisations to assess their current capability and then plot a roadmap for improvement prioritised by those KPAs which will make the biggest impact on performance.

White paper	A white paper is an authoritative report or guide that helps solve a problem. White papers are used to educate readers and help people make decisions, and may be a consultation as to the details of new legislation. The publishing of a white paper signifies a clear intention on the part of a government to pass new law.
Business plan	A business plan is a formal statement of a set of business goals, the reasons they are believed attainable, and the plan for reaching those goals. It may also contain background information about the organization or team attempting to reach those goals. Business plans may also target changes in perception and branding by the customer, client, taxpayer, or larger community.
Categories	On May 14, 1867, the 27-year-old Charles Sanders Peirce, who eventually founded Pragmatism, presented a paper entitled 'On a New List of Categories' to the American Academy of Arts and Sciences. Among other things, this paper outlined a theory of predication involving three universal categories that Peirce continued to apply in philosophy and elsewhere for the rest of his life. In the categories one will discern, concentrated, the pattern which one finds formed by the three grades of clearness in 'How to Make Our Ideas Clear' (1878 foundational paper for pragmatism), and in numerous other three-way distinctions in his work.
Exit strategy	An exit strategy is a means of leaving one's current situation, either after a predetermined objective has been achieved, or as a strategy to mitigate failure. An organisation or individual without an exit strategy may be in a quagmire. At worst, an exit strategy will save face; at best, an exit strategy will peg a withdrawal to the achievement of an objective worth more than the cost of continued involvement.
Marketing Intelligence	Marketing Intelligence is the information relevant to a company's markets, gathered and analyzed specifically for the purpose of accurate and confident decision-making in determining market opportunity, market penetration strategy, and market development metrics. Marketing intelligence is necessary when entering a foreign market. Marketing Intelligence is not the same as Market Intelligence (MARKINT).
Marketing strategy	Marketing strategy is a process that can allow an organization to concentrate its limited resources on the greatest opportunities to increase sales and achieve a sustainable competitive advantage. Marketing strategies serve as the fundamental underpinning of marketing plans designed to fill market needs and reach marketing objectives. Plans and objectives are generally tested for measurable results.
Mission statement	A mission statement is a statement of the purpose of a company or organization.

Chapter 13. Planning Reports and Proposals

	The mission statement should guide the actions of the organization, spell out its overall goal, provide a path, and guide decision-making. It provides 'the framework or context within which the company's strategies are formulated.' Effective mission statements start by effectively articulating the organization's purpose, its raison d'etre or reason for existing.
Content	In mathematics, a content is a real function μ defined on a field of sets $\mathcal{A}$ such that• $$\mu(A) \in [0, \infty] \text{ whenever } A \in \mathcal{A}..\mu(\emptyset) = 0..$$ $$\mu(A_1 \cup A_2) = \mu(A_1) + \mu(A_2) \text{ whenever } A_1, A_2 \in \mathcal{A} \text{ and } A_1 \cap A_2 = \emptyset.$$ A very important type of content is a measure, which is a σ-additive content defined on a σ-field. Every measure is a content, but not vice-versa.
Information architecture	Information architecture is the art and science of organizing and labelling websites, intranets, online communities and software to support usability. It is an emerging discipline and community of practice focused on bringing together principles of design and architecture to the digital landscape. Typically it involves a model or concept of information which is used and applied to activities that require explicit details of complex information systems.
Social media	Social media includes web- and mobile-based technologies which are used to turn communication into interactive dialogue among organizations, communities, and individuals. Andreas Kaplan and Michael Haenlein define social media as 'a group of Internet-based applications that build on the ideological and technological foundations of Web 2.0, and that allow the creation and exchange of user-generated content.' When the technologies are in place, social media is ubiquitously accessible, and enabled by scalable communication techniques. Social media technologies take on many different forms including magazines, Internet forums, weblogs, social blogs, microblogging, wikis, social networks, podcasts, photographs or pictures, video, rating and social bookmarking.
Virtual team	A virtual team is a group of individuals who work across time, space and organizational boundaries with links strengthened by webs of communication technology' Ale Ebrahim, N., Ahmed, S. & Taha, Z. in their recent (2009) literature review paper, added two key issues to definition of a virtual team 'as small temporary groups of geographically, organizationally and/ or time dispersed knowledge workers who coordinate their work predominantly with electronic information and communication technologies in order to accomplish one or more organization tasks' Members of virtual teams communicate electronically and may never meet face-to-face. Virtual teams are made possible by a proliferation of fiber optic technology that has significantly increased the scope of off-site communication.

Administration	As a legal concept, administration is a procedure under the insolvency laws of a number of common law jurisdictions. It functions as a rescue mechanism for insolvent entities and allows them to carry on running their business. The process - an alternative to liquidation - is often known as going into administration.
Small business	What constitutes a small business varies widely around the world. Small businesses are normally privately owned corporations, partnerships, or sole proprietorships. What constitutes 'small' in terms of government support and tax policy varies by country and by industry, ranging from fewer than 15 employees under the Australian Fair Work Act 2009, 50 employees according to the definition used by the European Union, and fewer than 500 employees to qualify for many U.S. Small Business Administration programs, although in 2006 there were over 18,000 'small businesses' with over 500 employees that accounted for half of all the employees employed by all 'small business '.
Chunking	In mathematics education at primary school level, chunking (sometimes also called the partial quotients method) is an elementary approach for solving simple division questions, by repeated subtraction. To calculate the result of dividing a large number by a small number, the student repeatedly takes away 'chunks' of the large number, where each 'chunk' is an easy multiple (for example 100×, 10×, 5× 2×, etc). of the small number, until the large number has been reduced to zero or the remainder is less than the divisor.
Discovery	Discovery is the act of detecting something new, or something 'old' that had been unknown. With reference to science and academic disciplines, discovery is the observation of new phenomena, new actions, or new events and providing new reasoning to explain the knowledge gathered through such observations with previously acquired knowledge from abstract thought and everyday experiences. Visual discoveries are often called sightings.
Due diligence	'Due diligence' is a term used for a number of concepts involving either an investigation of a business or person prior to signing a contract, or an act with a certain standard of care. It can be a legal obligation, but the term will more commonly apply to voluntary investigations. A common example of due diligence in various industries is the process through which a potential acquirer evaluates a target company or its assets for acquisition.
Failure analysis	Failure analysis is the process of collecting and analyzing data to determine the cause of a failure. It is an important discipline in many branches of manufacturing industry, such as the electronics industry, where it is a vital tool used in the development of new products and for the improvement of existing products.

Chapter 13. Planning Reports and Proposals

Market analysis	A market analysis studies the attractiveness and the dynamics of a special market within a special industry. It is part of the industry analysis and this in turn of the global environmental analysis. Through all these analyses the opportunities, strengths, weaknesses and threats of a company can be identified.
Request for proposal	A request for proposal is issued at an early stage in a procurement process, where an invitation is presented for suppliers, often through a bidding process, to submit a proposal on a specific commodity or service. The Request for proposal process brings structure to the procurement decision and is meant to allow the risks and benefits to be identified clearly up front. The Request for proposal may dictate to varying degrees the exact structure and format of the supplier's response.
Sales pitch	In selling technique, a sales pitch is a line of talk that attempts to persuade someone or something, with a planned sales presentation strategy of a product or service designed to initiate and close a sale of the product or service. A sales pitch is a planned presentation of a product or service designed to initiate and close a sale of the same product or service. A sales pitch is essentially designed to be either an introduction of a product or service to an audience who knows nothing about it, or a descriptive expansion of a product or service that an audience has already expressed interest in.
Credential	A credential is an attestation of qualification, competence, or authority issued to an individual by a third party with a relevant or de facto authority or assumed competence to do so. Examples of credentials include academic diplomas, academic degrees, certifications, security clearances, identification documents, badges, passwords, user names, keys, powers of attorney, and so on. Sometimes publications, such as scientific papers or books, may be viewed as similar to credentials by some people, especially if the publication was peer reviewed or made in a well-known journal or reputable publisher.
Annual report	An annual report is a comprehensive report on a company's activities throughout the preceding year. Annual reports are intended to give shareholders and other interested people information about the company's activities and financial performance. Most jurisdictions require companies to prepare and disclose annual reports, and many require the annual report to be filed at the company's registry.

1. _____ is the information relevant to a company's markets, gathered and analyzed specifically for the purpose of accurate and confident decision-making in determining market opportunity, market penetration strategy, and market development metrics. _____ is necessary when entering a foreign market. _____ is not the same as Market Intelligence (MARKINT).

   a. Marketing management
   b. Marketing mix
   c. Marketing mix for product software
   d. Marketing Intelligence

2. A _____ is a process in which a potential employee is evaluated by an employer for prospective employment in their company, organization, or firm. During this process, the employer hopes to determine whether or not the applicant is suitable for the role.

   A _____ typically precedes the hiring decision, and is used to evaluate the candidate.

   a. Job interview
   b. Multifactor
   c. Simple interest
   d. Gramm-Leach-Bliley Act

3. In logic a _____ is either (a) a meaningful declarative sentence that is either true or false, or (b) what is asserted or made by the use of a declarative sentence. In the latter case, a _____ is distinct from a sentence in that a sentence is only one formulation of a _____, whereas there may be many other formulations expressing the same _____.

   Philosopher of language, Peter Strawson advocated the use of the term '_____' in sense (b) in preference to proposition.

   a. Tacit assumption
   b. Term logic
   c. Testability
   d. Statement

4. _____, Programme and Project Management Maturity Model is a reference guide for structured best practice. It breaks down the broad disciplines of portfolio, programme and project management into a hierarchy of Key Process Areas (KPAs). The hierarchical approach enables organisations to assess their current capability and then plot a roadmap for improvement prioritised by those KPAs which will make the biggest impact on performance.

   a. P3M3
   b. PM Declaration of Interdependence
   c. Pmhub
   d. Point of total assumption

5. _____ is the act of detecting something new, or something 'old' that had been unknown. With reference to science and academic disciplines, _____ is the observation of new phenomena, new actions, or new events and providing new reasoning to explain the knowledge gathered through such observations with previously acquired knowledge from abstract thought and everyday experiences. Visual _____(ies) are often called sightings.

a. Discrimination learning
b. Dreyfus model of skill acquisition
c. Dysrationalia
d. Discovery

1. d
2. a
3. d
4. a
5. d

You can take the complete Chapter Practice Test

for Chapter 13. Planning Reports and Proposals
on all key terms, persons, places, and concepts.

Online 99 Cents

http://www.epub2.4.21191.13.cram101.com/

Use www.Cram101.com for all your study needs

including Cram101's online interactive problem solving labs in

chemistry, statistics, mathematics, and more.

CHAPTER OUTLINE: KEY TERMS, PEOPLE, PLACES, CONCEPTS

Annual report

Information needs

Proposal

Statement

Credential

Cultural identity

Intercultural

Intercultural communication

Storyboard

Content

Job interview

Public opinion

Social networking

Independent contractor

Request for proposal

Statement of work

Discovery

P3M3

Social media

_____ | Content marketing_____

_____ | Marketing_____

_____ | Law firm_____

_____ | International trade_____

Annual report	An annual report is a comprehensive report on a company's activities throughout the preceding year. Annual reports are intended to give shareholders and other interested people information about the company's activities and financial performance. Most jurisdictions require companies to prepare and disclose annual reports, and many require the annual report to be filed at the company's registry.
Information needs	Information need is an individual or group's desire to locate and obtain information to satisfy a conscious or unconscious need. The 'information' and 'need' in 'information need' are inseparable interconnection. Needs and interests call forth information.
Proposal	A business proposal is a written offer from a seller to a prospective buyer. Business proposals are often a key step in the complex sales process--i.e., whenever a buyer considers more than price in a purchase. There are three distinct categories of business proposals:•formally solicited•informally solicited•unsolicited. Solicited proposals are written in response to published requirements, contained in a Request for Proposal (RFP), Request for Quotation (RFQ), Request for Information (RFI) or an Invitation For Bid (IFB).
Statement	In logic a statement is either (a) a meaningful declarative sentence that is either true or false, or (b) what is asserted or made by the use of a declarative sentence. In the latter case, a statement is distinct from a sentence in that a sentence is only one formulation of a statement, whereas there may be many other formulations expressing the same statement.

Credential	A credential is an attestation of qualification, competence, or authority issued to an individual by a third party with a relevant or de facto authority or assumed competence to do so.
	Examples of credentials include academic diplomas, academic degrees, certifications, security clearances, identification documents, badges, passwords, user names, keys, powers of attorney, and so on. Sometimes publications, such as scientific papers or books, may be viewed as similar to credentials by some people, especially if the publication was peer reviewed or made in a well-known journal or reputable publisher.
Cultural identity	Cultural identity is the identity of a group or culture, or of an individual as far as one is influenced by one's belonging to a group or culture. Cultural identity is similar to and has overlaps with, but is not synonymous with, identity politics. Description
	Various modern cultural studies and social theories have investigated cultural identity.
Intercultural	Cross-cultural communication (also frequently referred to as intercultural communication, which is also used in a different sense, though) is a field of study that looks at how people from differing cultural backgrounds communicate, in similar and different ways among themselves, and how they endeavour to communicate across cultures. Origins
	During the Cold War, the United States economy was largely self-contained because the world was polarized into two separate and competing powers: the east and west. However, changes and advancements in economic relationships, political systems, and technological options began to break down old cultural barriers.
Intercultural communication	Intercultural communication is a form of global communication. It is used to describe the wide range of communication problems that naturally appear within an organization made up of individuals from different religious, social, ethnic, and educational backgrounds. Intercultural communication is sometimes used synonymously with cross-cultural communication.
Storyboard	Storyboards are graphic organizers in the form of illustrations or images displayed in sequence for the purpose of pre-visualizing a motion picture, animation, motion graphic or interactive media sequence.
	The storyboarding process, in the form it is known today, was developed at the Walt Disney Studio during the early 1930s, after several years of similar processes being in use at Walt Disney and other animation studios. Origins
	The storyboarding process can be very time-consuming and intricate.
Content	In mathematics, a content is a real function μ defined on a field of sets $\mathcal{A}$ such that·

Chapter 14. Writing Reports and Proposals

$$\mu(A) \in [0, \infty] \text{ whenever } A \in \mathcal{A}., \mu(\varnothing) = 0.,$$
$$\mu(A_1 \cup A_2) = \mu(A_1) + \mu(A_2) \text{ whenever } A_1, A_2 \in \mathcal{A} \text{ and } A_1 \cap A_2 = \varnothing.$$

A very important type of content is a measure, which is a σ-additive content defined on a σ-field. Every measure is a content, but not vice-versa.

Job interview	A job interview is a process in which a potential employee is evaluated by an employer for prospective employment in their company, organization, or firm. During this process, the employer hopes to determine whether or not the applicant is suitable for the role.
	A job interview typically precedes the hiring decision, and is used to evaluate the candidate.
Public opinion	Public opinion is the aggregate of individual attitudes or beliefs held by the adult population. Public opinion can also be defined as the complex collection of opinions of many different people and the sum of all their views.
	The principle approaches to the study of public opinion may be divided into 4 categories:•quantitative measurement of opinion distributions;•investigation of the internal relationships among the individual opinions that make up public opinion on an issue;•description or analysis of the public role of public opinion;•study both of the communication media that disseminate the ideas on which opinions are based and of the uses that propagandists and other manipulators make of these media.Concepts of public opinion
	Public opinion as a concept gained credence with the rise of 'public' in the eighteenth century.
Social networking	A social networking service is an online service, platform, or site that focuses on facilitating the building of social networks or social relations among people who, for example, share interests, activities, backgrounds, or real-life connections. A social network service consists of a representation of each user (often a profile), his/her social links, and a variety of additional services. Most social network services are web-based and provide means for users to interact over the Internet, such as e-mail and instant messaging.
Independent contractor	An independent contractor is a natural person, business, or corporation that provides goods or services to another entity under terms specified in a contract or within a verbal agreement. Unlike an employee, an independent contractor does not work regularly for an employer but works as and when required, during which time he or she may be subject to the Law of Agency. Independent contractors are usually paid on a freelance basis.
Request for proposal	A request for proposal is issued at an early stage in a procurement process, where an invitation is presented for suppliers, often through a bidding process, to submit a proposal on a specific commodity or service.

The Request for proposal process brings structure to the procurement decision and is meant to allow the risks and benefits to be identified clearly up front.

The Request for proposal may dictate to varying degrees the exact structure and format of the supplier's response.

Statement of work	A statement of work is a formal document that captures and defines the work activities, deliverables, and timeline a vendor must execute in performance of specified work for a client. The statement of work usually includes detailed requirements and pricing, with standard regulatory and governance terms and conditions.

Many formats and styles of Statement of Work document templates have been specialized for the hardware or software solutions described in the Request for Proposal. |
| Discovery | Discovery is the act of detecting something new, or something 'old' that had been unknown. With reference to science and academic disciplines, discovery is the observation of new phenomena, new actions, or new events and providing new reasoning to explain the knowledge gathered through such observations with previously acquired knowledge from abstract thought and everyday experiences. Visual discoveries are often called sightings. |
| P3M3 | P3M3, Programme and Project Management Maturity Model is a reference guide for structured best practice. It breaks down the broad disciplines of portfolio, programme and project management into a hierarchy of Key Process Areas (KPAs). The hierarchical approach enables organisations to assess their current capability and then plot a roadmap for improvement prioritised by those KPAs which will make the biggest impact on performance. |
| Social media | Social media includes web- and mobile-based technologies which are used to turn communication into interactive dialogue among organizations, communities, and individuals. Andreas Kaplan and Michael Haenlein define social media as 'a group of Internet-based applications that build on the ideological and technological foundations of Web 2.0, and that allow the creation and exchange of user-generated content.' When the technologies are in place, social media is ubiquitously accessible, and enabled by scalable communication techniques.

Social media technologies take on many different forms including magazines, Internet forums, weblogs, social blogs, microblogging, wikis, social networks, podcasts, photographs or pictures, video, rating and social bookmarking. |
| Content marketing | Content marketing is an umbrella term encompassing all marketing formats that involve the creation and sharing of content in order to engage current and potential consumer bases. Content marketing subscribes to the notion that delivering high-quality, relevant and valuable information to prospects and customers drives profitable consumer action. |

Chapter 14. Writing Reports and Proposals

Marketing	Marketing is 'the activity, set of institutions, and processes for creating, communicating, delivering, and exchanging offerings that have value for customers, clients, partners, and society at large.' For business to consumer marketing, it is 'the process by which companies create value for customers and build strong customer relationships, in order to capture value from customers in return'. For business to business marketing it is creating value, solutions, and relationships either short term or long term with a company or brand. It generates the strategy that underlies sales techniques, business communication, and business developments.
Law firm	A law firm is a business entity formed by one or more lawyers to engage in the practice of law. The primary service rendered by a law firm is to advise clients (individuals or corporations) about their legal rights and responsibilities, and to represent clients in civil or criminal cases, business transactions, and other matters in which legal advice and other assistance are sought. Law firms are organized in a variety of ways, depending on the jurisdiction in which the firm practices.
International trade	International trade is the exchange of capital, goods, and services across international borders or territories. In most countries, such trade represents a significant share of gross domestic product (GDP). While international trade has been present throughout much of history , its economic, social, and political importance has been on the rise in recent centuries.

1. _____ is the identity of a group or culture, or of an individual as far as one is influenced by one's belonging to a group or culture. _____ is similar to and has overlaps with, but is not synonymous with, identity politics. Description

   Various modern cultural studies and social theories have investigated _____.

   a. Jimmy Carter
   b. Democratization of knowledge
   c. Descriptive knowledge
   d. Cultural identity

2. . An _____ is a comprehensive report on a company's activities throughout the preceding year. _____s are intended to give shareholders and other interested people information about the company's activities and financial performance. Most jurisdictions require companies to prepare and disclose _____s, and many require the _____ to be filed at the company's registry.

a. Annual report

b. Alain Chartier

c. Alice Neville

d. Isabeau

3. A _____ is an attestation of qualification, competence, or authority issued to an individual by a third party with a relevant or de facto authority or assumed competence to do so.

Examples of _____s include academic diplomas, academic degrees, certifications, security clearances, identification documents, badges, passwords, user names, keys, powers of attorney, and so on. Sometimes publications, such as scientific papers or books, may be viewed as similar to _____s by some people, especially if the publication was peer reviewed or made in a well-known journal or reputable publisher.

a. Crossdisciplinarity

b. Credential

c. Descriptive knowledge

d. Dispersed knowledge

4. A _____ is a business entity formed by one or more lawyers to engage in the practice of law. The primary service rendered by a _____ is to advise clients (individuals or corporations) about their legal rights and responsibilities, and to represent clients in civil or criminal cases, business transactions, and other matters in which legal advice and other assistance are sought.

_____s are organized in a variety of ways, depending on the jurisdiction in which the firm practices.

a. Legal personality

b. Limited company

c. Limited liability limited partnership

d. Law firm

5. A _____ is issued at an early stage in a procurement process, where an invitation is presented for suppliers, often through a bidding process, to submit a proposal on a specific commodity or service. The _____ process brings structure to the procurement decision and is meant to allow the risks and benefits to be identified clearly up front.

The _____ may dictate to varying degrees the exact structure and format of the supplier's response.

a. Supplier diversity

b. Procurement outsourcing

c. Request for proposal

d. Integrity Inventory

1. d
2. a
3. b
4. d
5. c

You can take the complete Chapter Practice Test

for Chapter 14. Writing Reports and Proposals
on all key terms, persons, places, and concepts.

Online 99 Cents

http://www.epub2.4.21191.14.cram101.com/

Use www.Cram101.com for all your study needs

including Cram101's online interactive problem solving labs in

chemistry, statistics, mathematics, and more.

Chapter 15. Completing Reports and Proposals

_____ | Entrepreneur

_____ | Executive summary

_____ | Formality

_____ | Proposal

_____ | Financial Reporting

_____ | Independent contractor

_____ | Content

_____ | Intercultural

_____ | Intercultural communication

_____ | Storyboard

_____ | Cultural identity

_____ | Numbering

_____ | Receiver

_____ | Request for production

_____ | Request for proposal

_____ | Job interview

_____ | Social networking

_____ | Content management

_____ | Content management system

CHAPTER OUTLINE: KEY TERMS, PEOPLE, PLACES, CONCEPTS

_____ | Management system

_____ | Web content management system

_____ | Franchising

CHAPTER HIGHLIGHTS & NOTES: KEY TERMS, PEOPLE, PLACES, CONCEPTS

Entrepreneur	An entrepreneur is an owner or manager of a business enterprise who makes money through risk and/or initiative. The term was originally a loanword from French and was first defined by the Irish-French economist Richard Cantillon. Entrepreneur in English is a term applied to a person who is willing to help launch a new venture or enterprise and accept full responsibility for the outcome.
Executive summary	An executive summary, is a short document or section of a document, produced for business purposes, that summarizes a longer report or proposal or a group of related reports, in such a way that readers can rapidly become acquainted with a large body of material without having to read it all. It will usually contain a brief statement of the problem or proposal covered in the major document(s), background information, concise analysis and main conclusions. It is intended as an aid to decision making by managers and has been described as possibly the most important part of a business plan.
Formality	Utterances, conceptually similar to a ritual although typically secular and less involved. A formality may be as simple as a handshake upon making new acquaintances in Western culture to the carefully defined procedure of bows, handshakes, formal greetings, and business-card exchanges that may mark two businessmen being introduced in Japan. In legal and diplomatic circles, formalities include such matters as greeting an arriving head of state with the appropriate national anthem.
Proposal	A business proposal is a written offer from a seller to a prospective buyer. Business proposals are often a key step in the complex sales process--i.e., whenever a buyer considers more than price in a purchase. There are three distinct categories of business proposals:•formally solicited•informally solicited•unsolicited.

Chapter 15. Completing Reports and Proposals

Financial Reporting	Financial reporting is the process of preparing and distributing financial information to users of such information in various forms. The most common format of formal financial reporting are financial statements. Financial statements are prepared in accordance with rigorously applied standards defined by professional accounting bodies developed according to the legal and professional framework of a specific locale.
Independent contractor	An independent contractor is a natural person, business, or corporation that provides goods or services to another entity under terms specified in a contract or within a verbal agreement. Unlike an employee, an independent contractor does not work regularly for an employer but works as and when required, during which time he or she may be subject to the Law of Agency. Independent contractors are usually paid on a freelance basis.
Content	In mathematics, a content is a real function μ defined on a field of sets $\mathcal{A}$ such that• $$\mu(A) \in [0, \infty] \text{ whenever } A \in \mathcal{A}.\text{, }\mu(\varnothing) = 0.\text{,}$$ $$\mu(A_1 \cup A_2) = \mu(A_1) + \mu(A_2) \text{ whenever } A_1, A_2 \in \mathcal{A} \text{ and } A_1 \cap A_2 = \varnothing.$$ A very important type of content is a measure, which is a σ-additive content defined on a σ-field. Every measure is a content, but not vice-versa.
Intercultural	Cross-cultural communication (also frequently referred to as intercultural communication, which is also used in a different sense, though) is a field of study that looks at how people from differing cultural backgrounds communicate, in similar and different ways among themselves, and how they endeavour to communicate across cultures. Origins During the Cold War, the United States economy was largely self-contained because the world was polarized into two separate and competing powers: the east and west. However, changes and advancements in economic relationships, political systems, and technological options began to break down old cultural barriers.
Intercultural communication	Intercultural communication is a form of global communication. It is used to describe the wide range of communication problems that naturally appear within an organization made up of individuals from different religious, social, ethnic, and educational backgrounds. Intercultural communication is sometimes used synonymously with cross-cultural communication.
Storyboard	Storyboards are graphic organizers in the form of illustrations or images displayed in sequence for the purpose of pre-visualizing a motion picture, animation, motion graphic or interactive media sequence. The storyboarding process, in the form it is known today, was developed at the Walt Disney Studio during the early 1930s, after several years of similar processes being in use at Walt Disney and other animation studios. Origins

Cultural identity	Cultural identity is the identity of a group or culture, or of an individual as far as one is influenced by one's belonging to a group or culture. Cultural identity is similar to and has overlaps with, but is not synonymous with, identity politics. Description
	Various modern cultural studies and social theories have investigated cultural identity.
Numbering	In computability theory a numbering is the assignment of natural numbers to a set of objects like rational numbers, graphs or words in some language. A numbering can be used to transfer the idea of computability and related concepts, which are strictly defined on the natural numbers using computable functions, to different objects.
	Important numberings are the Gödel numbering of the terms in first-order predicate calculus and numberings of the set of computable functions which can be used to apply results of computability theory on the set of computable functions itself.
Receiver	In modulated ultrasound terminology, a receiver is a device that receives a modulated ultrasound signal and decodes it for use as sound, navigational-position information, etc. Its function is somewhat like that of a radio receiver
Request for production	A request for production is a legal request for documents, electronically stored information, or other tangible items. In civil procedure, during the discovery phase of litigation, a party to a lawsuit may request that another party provide any documents that it has that pertain to the subject matter of the lawsuit. For example, a party in a court case may obtain copies of e-mail messages sent by employees of the opposing party.
Request for proposal	A request for proposal is issued at an early stage in a procurement process, where an invitation is presented for suppliers, often through a bidding process, to submit a proposal on a specific commodity or service. The Request for proposal process brings structure to the procurement decision and is meant to allow the risks and benefits to be identified clearly up front.
	The Request for proposal may dictate to varying degrees the exact structure and format of the supplier's response.
Job interview	A job interview is a process in which a potential employee is evaluated by an employer for prospective employment in their company, organization, or firm. During this process, the employer hopes to determine whether or not the applicant is suitable for the role.
	A job interview typically precedes the hiring decision, and is used to evaluate the candidate.

Chapter 15. Completing Reports and Proposals

Social networking	A social networking service is an online service, platform, or site that focuses on facilitating the building of social networks or social relations among people who, for example, share interests, activities, backgrounds, or real-life connections. A social network service consists of a representation of each user (often a profile), his/her social links, and a variety of additional services. Most social network services are web-based and provide means for users to interact over the Internet, such as e-mail and instant messaging.
Content management	Content management, is the set of processes and technologies that support the collection, managing, and publishing of information in any form or medium. In recent times this information is typically referred to as content or, to be precise, digital content. Digital content may take the form of text (such as electronic documents), multimedia files (such as audio or video files), or any other file type that follows a content lifecycle requiring management.
Content management system	A content management system is a computer system that allows publishing, editing, and modifying content as well as site maintenance from a central page. It provides a collection of procedures used to manage workflow in a collaborative environment. These procedures can be manual or computer-based.
Management system	A management system is the framework of processes and procedures used to ensure that an organization can fulfill all tasks required to achieve its objectives. For instance, an environmental management system enables organizations to improve their environmental performance through a process of continuous improvement. An oversimplification is 'Plan, Do, Check, Act'.
Web content management system	A Web Content Management System is a software system that provides website authoring, collaboration, and administration tools designed to allow users with little knowledge of web programming languages or markup languages to create and manage website content with relative ease. A robust provides the foundation for collaboration, offering users the ability to manage documents and output for multiple author editing and participation. Most systems use a Content Repository or a database to store page content, metadata, and other information assets that might be needed by the system.
Franchising	Franchising is the practice of using another firm's successful business model. The word 'franchise' is of anglo-French derivation - from franc - meaning free, and is used both as a noun and as a (transitive) verb. For the franchisor, the franchise is an alternative to building 'chain stores' to distribute goods that avoids the investments and liability of a chain.

1. A _____ is issued at an early stage in a procurement process, where an invitation is presented for suppliers, often through a bidding process, to submit a proposal on a specific commodity or service. The _____ process brings structure to the procurement decision and is meant to allow the risks and benefits to be identified clearly up front.

   The _____ may dictate to varying degrees the exact structure and format of the supplier's response.

   a. Supplier diversity
   b. Procurement outsourcing
   c. Jimmy Carter
   d. Request for proposal

2. An _____ is a natural person, business, or corporation that provides goods or services to another entity under terms specified in a contract or within a verbal agreement. Unlike an employee, an _____ does not work regularly for an employer but works as and when required, during which time he or she may be subject to the Law of Agency. _____s are usually paid on a freelance basis.

   a. Independent contractor
   b. Induction programme
   c. INGRADA
   d. Integrity Inventory

3. A business _____ is a written offer from a seller to a prospective buyer. Business _____s are often a key step in the complex sales process--i.e., whenever a buyer considers more than price in a purchase.

   There are three distinct categories of business _____s:•formally solicited•informally solicited•unsolicited.

   Solicited _____s are written in response to published requirements, contained in a Request for _____ (RFP), Request for Quotation (RFQ), Request for Information (RFI) or an Invitation For Bid (IFB).

   a. Qualified prospect
   b. Quosal
   c. Proposal
   d. Sale and rent back

4. . _____, is the set of processes and technologies that support the collection, managing, and publishing of information in any form or medium. In recent times this information is typically referred to as content or, to be precise, digital content. Digital content may take the form of text (such as electronic documents), multimedia files (such as audio or video files), or any other file type that follows a content lifecycle requiring management.

   a. DMAIC
   b. data security
   c. Content management

5. An _____ is an owner or manager of a business enterprise who makes money through risk and/or initiative. The term was originally a loanword from French and was first defined by the Irish-French economist Richard Cantillon. _____ in English is a term applied to a person who is willing to help launch a new venture or enterprise and accept full responsibility for the outcome.

    a. Entrepreneur
    b. Entrepreneurial ecosystem
    c. Entrepreneurial finance
    d. Entrepreneurial Management Center

1. d
2. a
3. c
4. c
5. a

You can take the complete Chapter Practice Test

for Chapter 15. Completing Reports and Proposals
on all key terms, persons, places, and concepts.

Online 99 Cents

http://www.epub2.4.21191.15.cram101.com/

Use www.Cram101.com for all your study needs

including Cram101's online interactive problem solving labs in

chemistry, statistics, mathematics, and more.

CHAPTER OUTLINE: KEY TERMS, PEOPLE, PLACES, CONCEPTS

Cultural identity

Negotiation

Audience analysis

Public opinion

Baby boomer

Discovery

Screencast

Social media

Attention span

Search Engine

Independent contractor

Storyboard

Orientation

Formality

Credential

Job interview

Proposal

Action item

Delivery

	Impromptu speaking
	Interactivity
	Online shopping
	Public speaking
	Sign language
	Get Satisfaction
	Self-assessment

CHAPTER HIGHLIGHTS & NOTES: KEY TERMS, PEOPLE, PLACES, CONCEPTS

Cultural identity	Cultural identity is the identity of a group or culture, or of an individual as far as one is influenced by one's belonging to a group or culture. Cultural identity is similar to and has overlaps with, but is not synonymous with, identity politics. Description
	Various modern cultural studies and social theories have investigated cultural identity.
Negotiation	Negotiation is a dialogue between two or more people or parties, intended to reach an understanding, resolve point of difference, or gain advantage in outcome of dialogue, to produce an agreement upon courses of action, to bargain for individual or collective advantage, to craft outcomes to satisfy various interests of two people/parties involved in negotiation process. Negotiation is a process where each party involved in negotiating tries to gain an advantage for themselves by the end of the process. Negotiation is intended to aim at compromise.
Audience analysis	Audience analysis is a task that is often performed by technical writers in a project's early stages. It consists of assessing the audience to make sure the information provided to them is at the appropriate level. The audience is often referred to as the end-user, and all communications need to be targeted towards the defined audience.
Public opinion	Public opinion is the aggregate of individual attitudes or beliefs held by the adult population.

Public opinion can also be defined as the complex collection of opinions of many different people and the sum of all their views.

The principle approaches to the study of public opinion may be divided into 4 categories:•quantitative measurement of opinion distributions;•investigation of the internal relationships among the individual opinions that make up public opinion on an issue;•description or analysis of the public role of public opinion;•study both of the communication media that disseminate the ideas on which opinions are based and of the uses that propagandists and other manipulators make of these media.Concepts of public opinion

Public opinion as a concept gained credence with the rise of 'public' in the eighteenth century.

Baby boomer	A baby boomer is a person who was born during the demographic Post-World War II baby boom between the years 1946 and 1964, according to the U.S. Census Bureau. The term 'baby boomer' is sometimes used in a cultural context. Therefore, it is impossible to achieve broad consensus of a precise definition, even within a given territory.
Discovery	Discovery is the act of detecting something new, or something 'old' that had been unknown. With reference to science and academic disciplines, discovery is the observation of new phenomena, new actions, or new events and providing new reasoning to explain the knowledge gathered through such observations with previously acquired knowledge from abstract thought and everyday experiences. Visual discoveries are often called sightings.
Screencast	A screencast is a digital recording of computer screen output, also known as a video screen capture, often containing audio narration. The term screencast compares with the related term screenshot; whereas screenshot is a picture of a computer screen, a screencast is essentially a movie of the changes over time that a user sees on a computer screen, enhanced with audio narration. In 2004, columnist Jon Udell invited readers of his blog to propose names for the emerging genre.
Social media	Social media includes web- and mobile-based technologies which are used to turn communication into interactive dialogue among organizations, communities, and individuals. Andreas Kaplan and Michael Haenlein define social media as 'a group of Internet-based applications that build on the ideological and technological foundations of Web 2.0, and that allow the creation and exchange of user-generated content.' When the technologies are in place, social media is ubiquitously accessible, and enabled by scalable communication techniques.

Chapter 16. Developing Oral and Online Presentations

Attention span	Attention span is the amount of time that a person can concentrate on a task without becoming distracted. Most educators and psychologists agree that the ability to focus one's attention on a task is crucial for the achievement of one's goals. Estimates for the length of human attention span are highly variable and depend on the precise definition of attention being used.
Search Engine	Search Engine was a weekly Canadian radio show that aired on CBC Radio One, then a dedicated podcast available on CBC.ca, and now a podcast on TVOntario's website, tvo.org. It is hosted by Jesse Brown, who also co-produces the show with Geoff Siskind and Andrew Parker. Cory Doctorow, novelist and editor of Boing Boing, is also a regular contributor.
Independent contractor	An independent contractor is a natural person, business, or corporation that provides goods or services to another entity under terms specified in a contract or within a verbal agreement. Unlike an employee, an independent contractor does not work regularly for an employer but works as and when required, during which time he or she may be subject to the Law of Agency. Independent contractors are usually paid on a freelance basis.
Storyboard	Storyboards are graphic organizers in the form of illustrations or images displayed in sequence for the purpose of pre-visualizing a motion picture, animation, motion graphic or interactive media sequence. The storyboarding process, in the form it is known today, was developed at the Walt Disney Studio during the early 1930s, after several years of similar processes being in use at Walt Disney and other animation studios. Origins The storyboarding process can be very time-consuming and intricate.
Orientation	In mathematics, orientation is a geometric notion that in two dimensions allows one to say when a cycle goes around clockwise or counterclockwise, and in three dimensions when a figure is left-handed or right-handed. In linear algebra, the notion of orientation makes sense in arbitrary dimensions. In this setting, the orientation of an ordered basis is a kind of asymmetry that makes a reflection impossible to replicate by means of a simple rotation.
Formality	Utterances, conceptually similar to a ritual although typically secular and less involved. A formality may be as simple as a handshake upon making new acquaintances in Western culture to the carefully defined procedure of bows, handshakes, formal greetings, and business-card exchanges that may mark two businessmen being introduced in Japan. In legal and diplomatic circles, formalities include such matters as greeting an arriving head of state with the appropriate national anthem.

Credential	A credential is an attestation of qualification, competence, or authority issued to an individual by a third party with a relevant or de facto authority or assumed competence to do so.
	Examples of credentials include academic diplomas, academic degrees, certifications, security clearances, identification documents, badges, passwords, user names, keys, powers of attorney, and so on. Sometimes publications, such as scientific papers or books, may be viewed as similar to credentials by some people, especially if the publication was peer reviewed or made in a well-known journal or reputable publisher.
Job interview	A job interview is a process in which a potential employee is evaluated by an employer for prospective employment in their company, organization, or firm. During this process, the employer hopes to determine whether or not the applicant is suitable for the role.
	A job interview typically precedes the hiring decision, and is used to evaluate the candidate.
Proposal	A business proposal is a written offer from a seller to a prospective buyer. Business proposals are often a key step in the complex sales process--i.e., whenever a buyer considers more than price in a purchase.
	There are three distinct categories of business proposals:•formally solicited•informally solicited•unsolicited.
	Solicited proposals are written in response to published requirements, contained in a Request for Proposal (RFP), Request for Quotation (RFQ), Request for Information (RFI) or an Invitation For Bid (IFB).
Action item	In management, an action item is a documented event, task, activity, or action that needs to take place. Action items are discrete units that can be handled by a single person.
	Action items are usually created during a discussion by a group of people who are meeting about one or more topics and during the discussion it is discovered that some kind of action is needed.
Delivery	Delivery is the process of transporting goods. Most goods are delivered through a transportation network. Cargo (physical goods) are primarily delivered via roads and railroads on land, shipping lanes on the sea and airline networks in the air.
Impromptu speaking	Impromptu speaking is a speech and debate consolation event that involves an eight minute speech, with up to three of these eight minutes available for use as preparation time (known as prep time, or simply prep).

Chapter 16. Developing Oral and Online Presentations

At the college level, the speaker is granted seven minutes to divide as he or she sees fit, as stipulated by the National Forensics Association and the American Forensics Association.

Another variation exists in which the speaker must speak for five minutes and half of a minute is given for preparation time.

Interactivity

In the fields of information science, communication, and industrial design, there is debate over the meaning of interactivity. In the 'contingency view' of interactivity, there are three levels:•Noninteractive, when a message is not related to previous messages;•Reactive, when a message is related only to one immediately previous message; and•Interactive, when a message is related to a number of previous messages and to the relationship between them. Human to human communication

Human communication is the basic example of interactive communication which involves two different processes; human to human interactivity and human to computer interactivity. Human-Human interactivity is the communication between people.

Online shopping

Online shopping is a form of electronic commerce whereby consumers directly buy goods or services from a seller over the Internet without an intermediary service. An online shop, eshop, e-store, Internet shop, webshop, webstore, online store, or virtual store evokes the physical analogy of buying products or services at a bricks-and-mortar retailer or shopping centre. The process is called business-to-consumer (B2C) online shopping.

Public speaking

Public speaking is the process of speaking to a group of people in a structured, deliberate manner intended to inform, influence, or entertain the listeners. It is closely allied to 'presenting', although the latter has more of a commercial advertisement.

In public speaking, as in any form of communication, there are five basic elements, often expressed as 'who is saying what to whom using what medium with what effects?' The purpose of public speaking can range from simply transmitting information, to motivating people to act, to simply telling a story.

Sign language

A sign language is a language which, instead of acoustically conveyed sound patterns, uses manual communication and body language to convey meaning. This can involve simultaneously combining hand shapes, orientation and movement of the hands, arms or body, and facial expressions to fluidly express a speaker's thoughts.

Wherever communities of deaf people exist, sign languages develop.

Get Satisfaction	Get Satisfaction is a customer community software platform for technical support based in San Francisco, California, United States. It was founded on January 31, 2007 by several people, including Lane Becker, Amy Muller, Thor Muller, and Jonathan Grubb. It publicly launched in September 2007. The idea for the service originated from Valleyschwag as a side project.
Self-assessment	In social psychology, self-assessment is the process of looking at oneself in order to assess aspects that are important to one's identity. It is one of the motives that drive self-evaluation, along with self-verification and self-enhancement. Sedikides (1993) suggests that the self-assessment motive will prompt people to seek information to confirm their uncertain self-concept rather than their certain self-concept and at the same time people use self-assessment to enhance their certainty of their own self-knowledge.

CHAPTER QUIZ: KEY TERMS, PEOPLE, PLACES, CONCEPTS

1. In social psychology, _____ is the process of looking at oneself in order to assess aspects that are important to one's identity. It is one of the motives that drive self-evaluation, along with self-verification and self-enhancement. Sedikides (1993) suggests that the _____ motive will prompt people to seek information to confirm their uncertain self-concept rather than their certain self-concept and at the same time people use _____ to enhance their certainty of their own self-knowledge.

    a. Proof of concept
    b. Questionnaire
    c. Factorial design
    d. Self-assessment

2. A _____ is a digital recording of computer screen output, also known as a video screen capture, often containing audio narration. The term _____ compares with the related term screenshot; whereas screenshot is a picture of a computer screen, a _____ is essentially a movie of the changes over time that a user sees on a computer screen, enhanced with audio narration.

    In 2004, columnist Jon Udell invited readers of his blog to propose names for the emerging genre.

    a. Simulation game
    b. Screencast
    c. Suspension training
    d. Talascend

Chapter 16. Developing Oral and Online Presentations

3. _____ includes web- and mobile-based technologies which are used to turn communication into interactive dialogue among organizations, communities, and individuals. Andreas Kaplan and Michael Haenlein define _____ as 'a group of Internet-based applications that build on the ideological and technological foundations of Web 2.0, and that allow the creation and exchange of user-generated content.' When the technologies are in place, _____ is ubiquitously accessible, and enabled by scalable communication techniques. Classification of _____

   _____ technologies take on many different forms including magazines, Internet forums, weblogs, social blogs, microblogging, wikis, social networks, podcasts, photographs or pictures, video, rating and social bookmarking.

   a. Jimmy Carter
   b. Strength and conditioning coach
   c. Social media
   d. Talascend

4. _____ was a weekly Canadian radio show that aired on CBC Radio One, then a dedicated podcast available on CBC.ca, and now a podcast on TVOntario's website, tvo.org. It is hosted by Jesse Brown, who also co-produces the show with Geoff Siskind and Andrew Parker. Cory Doctorow, novelist and editor of Boing Boing, is also a regular contributor.

   a. Security Now
   b. This Week in Photography
   c. Search Engine
   d. Tips from the Top Floor

5. _____ is the act of detecting something new, or something 'old' that had been unknown. With reference to science and academic disciplines, _____ is the observation of new phenomena, new actions, or new events and providing new reasoning to explain the knowledge gathered through such observations with previously acquired knowledge from abstract thought and everyday experiences. Visual _____(ies) are often called sightings.

   a. Discovery
   b. Dreyfus model of skill acquisition
   c. Dysrationalia
   d. Habituation

190

1. d
2. b
3. c
4. c
5. a

You can take the complete Chapter Practice Test

for Chapter 16. Developing Oral and Online Presentations
on all key terms, persons, places, and concepts.

Online 99 Cents

http://www.epub2.4.21191.16.cram101.com/

Use www.Cram101.com for all your study needs

including Cram101's online interactive problem solving labs in

chemistry, statistics, mathematics, and more.

Discovery

Brainstorming

Information product

Overhead

Screencast

Social network

Storyboard

Whiteboard

Interactivity

Social media

Baby boomer

Artwork

Content

Social networking

P3M3

Audience analysis

Nutrition

Pecha Kucha

Chapter 17. Enhancing Presentations with Slides and Other Visuals

Discovery	Discovery is the act of detecting something new, or something 'old' that had been unknown. With reference to science and academic disciplines, discovery is the observation of new phenomena, new actions, or new events and providing new reasoning to explain the knowledge gathered through such observations with previously acquired knowledge from abstract thought and everyday experiences. Visual discoveries are often called sightings.
Brainstorming	Brainstorming is a group or individual creativity technique by which efforts are made to find a conclusion for a specific problem by gathering a list of ideas spontaneously contributed by its member(s). The term was popularized by Alex Faickney Osborn in the 1963 book Applied Imagination. Osborn claimed that brainstorming was more effective than individuals working alone in generating ideas, although more recent research has questioned this conclusion.
Information product	An Information Product is any final product in the form of information that a person needs to have. This Information Product consists of several Information Element, which are located in the organizational value chain. To illustrate the concept of an IP, an example is shown of a bottleneck analysis in HR (by J. Willems 2008).
Overhead	In business, overhead or overhead expense refers to an ongoing expense of operating a business (also known as Operating Expenses - rent, gas/electricity, wages etc).. The term overhead is usually used to group expenses that are necessary to the continued functioning of the business but cannot be immediately associated with the products/services being offered (e.g., do not directly generate profits). Closely related accounting concepts are fixed costs versus variable costs and indirect costs versus direct costs.
Screencast	A screencast is a digital recording of computer screen output, also known as a video screen capture, often containing audio narration. The term screencast compares with the related term screenshot; whereas screenshot is a picture of a computer screen, a screencast is essentially a movie of the changes over time that a user sees on a computer screen, enhanced with audio narration. In 2004, columnist Jon Udell invited readers of his blog to propose names for the emerging genre.
Social network	A social network is a social structure made up of a set of actors (such as individuals or organizations) and the dyadic ties between these actors. The social network perspective provides a clear way of analyzing the structure of whole social entities. The study of these structures uses social network analysis to identify local and global patterns, locate influential entities, and examine network dynamics.

Storyboard	Storyboards are graphic organizers in the form of illustrations or images displayed in sequence for the purpose of pre-visualizing a motion picture, animation, motion graphic or interactive media sequence. The storyboarding process, in the form it is known today, was developed at the Walt Disney Studio during the early 1930s, after several years of similar processes being in use at Walt Disney and other animation studios. Origins The storyboarding process can be very time-consuming and intricate.
Whiteboard	A whiteboard is a name for any glossy, usually white surface for nonpermanent markings. Whiteboards are analogous to chalkboards, allowing rapid marking and erasing of markings on their surface. The popularity of whiteboards increased rapidly in the mid-1990s and they have become a fixture in many offices, meeting rooms, school classrooms, and other work environments.
Interactivity	In the fields of information science, communication, and industrial design, there is debate over the meaning of interactivity. In the 'contingency view' of interactivity, there are three levels:•Noninteractive, when a message is not related to previous messages;•Reactive, when a message is related only to one immediately previous message; and•Interactive, when a message is related to a number of previous messages and to the relationship between them. Human to human communication Human communication is the basic example of interactive communication which involves two different processes; human to human interactivity and human to computer interactivity. Human-Human interactivity is the communication between people.
Social media	Social media includes web- and mobile-based technologies which are used to turn communication into interactive dialogue among organizations, communities, and individuals. Andreas Kaplan and Michael Haenlein define social media as 'a group of Internet-based applications that build on the ideological and technological foundations of Web 2.0, and that allow the creation and exchange of user-generated content.' When the technologies are in place, social media is ubiquitously accessible, and enabled by scalable communication techniques. Social media technologies take on many different forms including magazines, Internet forums, weblogs, social blogs, microblogging, wikis, social networks, podcasts, photographs or pictures, video, rating and social bookmarking.
Baby boomer	A baby boomer is a person who was born during the demographic Post-World War II baby boom between the years 1946 and 1964, according to the U.S. Census Bureau. The term 'baby boomer' is sometimes used in a cultural context.

Chapter 17. Enhancing Presentations with Slides and Other Visuals

Artwork	Artwork (also known as art layout drawing) is a type of drawing that serves a graphical representation of an image for its reproduction onto a substrate via various processes, such as silkscreen, label making and other printing methods. Transfer of the image often involves a photographic process. Historically, some types of artworks were prepared on clear polyester film or similar media for strength, durability and dimensional stability.
Content	In mathematics, a content is a real function μ defined on a field of sets $\mathcal{A}$ such that· $$\mu(A) \in [0, \infty] \text{ whenever } A \in \mathcal{A}. \, \mu(\varnothing) = 0. \,$$ $$\mu(A_1 \cup A_2) = \mu(A_1) + \mu(A_2) \text{ whenever } A_1, A_2 \in \mathcal{A} \text{ and } A_1 \cap A_2 = \varnothing.$$ A very important type of content is a measure, which is a σ-additive content defined on a σ-field. Every measure is a content, but not vice-versa.
Social networking	A social networking service is an online service, platform, or site that focuses on facilitating the building of social networks or social relations among people who, for example, share interests, activities, backgrounds, or real-life connections. A social network service consists of a representation of each user (often a profile), his/her social links, and a variety of additional services. Most social network services are web-based and provide means for users to interact over the Internet, such as e-mail and instant messaging.
P3M3	P3M3, Programme and Project Management Maturity Model is a reference guide for structured best practice. It breaks down the broad disciplines of portfolio, programme and project management into a hierarchy of Key Process Areas (KPAs). The hierarchical approach enables organisations to assess their current capability and then plot a roadmap for improvement prioritised by those KPAs which will make the biggest impact on performance.
Audience analysis	Audience analysis is a task that is often performed by technical writers in a project's early stages. It consists of assessing the audience to make sure the information provided to them is at the appropriate level. The audience is often referred to as the end-user, and all communications need to be targeted towards the defined audience.
Nutrition	Nutrition is the provision, to cells and organisms, of the materials necessary (in the form of food) to support life. Many common health problems can be prevented or alleviated with a healthy diet. The diet of an organism is what it eats, which is largely determined by the perceived palatability of foods.
Pecha Kucha	Pecha Kucha is a presentation methodology in which 20 slides are shown for 20 seconds each (approx.

6' 40' in total), usually seen in a multiple-speaker event called a Pecha Kucha Night (PKN).

Pecha Kucha Night was devised in February 2003 by Astrid Klein and Mark Dytham of Tokyo's Klein-Dytham Architecture (KDa), as a way to attract people to Super Deluxe, their experimental event space in Roppongi. Pecha Kucha Night events consist of around a dozen presentations, each presenter having 20 slides, each shown for 20 seconds on a timer.

1. A _____ is a social structure made up of a set of actors (such as individuals or organizations) and the dyadic ties between these actors. The _____ perspective provides a clear way of analyzing the structure of whole social entities. The study of these structures uses _____ analysis to identify local and global patterns, locate influential entities, and examine network dynamics.

    a. Social network
    b. Strength and conditioning coach
    c. Suspension training
    d. Talascend

2. An _____ is any final product in the form of information that a person needs to have. This _____ consists of several Information Element, which are located in the organizational value chain. To illustrate the concept of an IP, an example is shown of a bottleneck analysis in HR (by J. Willems 2008).

    a. Information product
    b. Intangible asset finance
    c. Interim management
    d. Office broker

3. _____ is a task that is often performed by technical writers in a project's early stages. It consists of assessing the audience to make sure the information provided to them is at the appropriate level. The audience is often referred to as the end-user, and all communications need to be targeted towards the defined audience.

    a. EbXML Messaging Services
    b. Electronic performance support systems
    c. Institute of Scientific and Technical Communicators
    d. Audience analysis

4. A _____ is a digital recording of computer screen output, also known as a video screen capture, often containing audio narration. The term _____ compares with the related term screenshot; whereas screenshot is a picture of a computer screen, a _____ is essentially a movie of the changes over time that a user sees on a computer screen, enhanced with audio narration.

In 2004, columnist Jon Udell invited readers of his blog to propose names for the emerging genre.

   a. Screencast
   b. Strength and conditioning coach
   c. Suspension training
   d. Talascend

5. A _____ is a name for any glossy, usually white surface for nonpermanent markings. _____s are analogous to chalkboards, allowing rapid marking and erasing of markings on their surface. The popularity of _____s increased rapidly in the mid-1990s and they have become a fixture in many offices, meeting rooms, school classrooms, and other work environments.

   a. Whiteboard
   b. Multifactor
   c. Simple interest
   d. Talascend

1. a
2. a
3. d
4. a
5. a

You can take the complete Chapter Practice Test

for Chapter 17. Enhancing Presentations with Slides and Other Visuals
on all key terms, persons, places, and concepts.

Online 99 Cents

http://www.epub2.4.21191.17.cram101.com/

Use www.Cram101.com for all your study needs

including Cram101's online interactive problem solving labs in

chemistry, statistics, mathematics, and more.

CHAPTER OUTLINE: KEY TERMS, PEOPLE, PLACES, CONCEPTS

	Human resources
	Social media
	Social network
	Workforce management
	Proposal
	Quality
	Survey research
	Career counseling
	Volunteering
	Curriculum vitae
	Public opinion
	P3M3
	Financial Services
	Personal branding
	Storyboard
	Podcasts
	Work experience
	Appearance
	Request for production

_____ | Production _____

_____ | Text file _____

_____ | Content _____

CHAPTER HIGHLIGHTS & NOTES: KEY TERMS, PEOPLE, PLACES, CONCEPTS

Human resources	Human resources is the set of individuals who make up the workforce of an organization, business sector or an economy. 'Human capital' is sometimes used synonymously with human resources, although human capital typically refers to a more narrow view; i.e., the knowledge the individuals embody and can contribute to an organization. Likewise, other terms sometimes used include 'manpower', 'talent', 'labor' or simply 'people'.
Social media	Social media includes web- and mobile-based technologies which are used to turn communication into interactive dialogue among organizations, communities, and individuals. Andreas Kaplan and Michael Haenlein define social media as 'a group of Internet-based applications that build on the ideological and technological foundations of Web 2.0, and that allow the creation and exchange of user-generated content.' When the technologies are in place, social media is ubiquitously accessible, and enabled by scalable communication techniques.
	Social media technologies take on many different forms including magazines, Internet forums, weblogs, social blogs, microblogging, wikis, social networks, podcasts, photographs or pictures, video, rating and social bookmarking.
Social network	A social network is a social structure made up of a set of actors (such as individuals or organizations) and the dyadic ties between these actors. The social network perspective provides a clear way of analyzing the structure of whole social entities. The study of these structures uses social network analysis to identify local and global patterns, locate influential entities, and examine network dynamics.
Workforce management	Workforce management encompasses all the activities needed to maintain a productive workforce. Under the umbrella of human resource management, WFM is sometimes referred to as HRMS systems, or even part of ERP systems.

Proposal	A business proposal is a written offer from a seller to a prospective buyer. Business proposals are often a key step in the complex sales process--i.e., whenever a buyer considers more than price in a purchase.
	There are three distinct categories of business proposals:•formally solicited•informally solicited•unsolicited.
	Solicited proposals are written in response to published requirements, contained in a Request for Proposal (RFP), Request for Quotation (RFQ), Request for Information (RFI) or an Invitation For Bid (IFB).
Quality	Quality in business, engineering and manufacturing has a pragmatic interpretation as the non-inferiority or superiority of something; it is also defined as fitness for purpose. Quality is a perceptual, conditional and somewhat subjective attribute and may be understood differently by different people. Consumers may focus on the specification quality of a product/service, or how it compares to competitors in the marketplace.
Survey research	Survey research involves utilizing interviews or questionnaires to obtain quantitative information in fields such as marketing, politics, and social science. Utilizing surveys is considered to be an efficient way of collecting data from a large number of respondents, accurately representing a whole population. Surveys also have the benefit of providing data that is relatively free from errors.
Career counseling	Career counseling and career coaching are similar in nature to traditional counseling. However, the focus is generally on issues such as career exploration, career change, personal career development and other career related issues. Typically when people come for career counseling they know exactly what they want to get out of the process, but are unsure about how it may work.
Volunteering	Volunteering is generally considered an altruistic activity, intended to promote good or improve human quality of life. It is considered as serving the society through own interest, personal skills or learning, which in return produces a feeling of self-worth and respect, instead of money. Volunteering is also famous for the skill development, to socialize and to have fun.
Curriculum vitae	A curriculum vitae (also spelled curriculum vitæ), provides an overview of a person's experience and other qualifications. In some countries, a CV is typically the first item that a potential employer encounters regarding the job seeker and is typically used to screen applicants, often followed by an interview, when seeking employment.

Chapter 18. Building Careers and Writing Résumés

Public opinion	Public opinion is the aggregate of individual attitudes or beliefs held by the adult population. Public opinion can also be defined as the complex collection of opinions of many different people and the sum of all their views. The principle approaches to the study of public opinion may be divided into 4 categories:•quantitative measurement of opinion distributions;•investigation of the internal relationships among the individual opinions that make up public opinion on an issue;•description or analysis of the public role of public opinion;•study both of the communication media that disseminate the ideas on which opinions are based and of the uses that propagandists and other manipulators make of these media.Concepts of public opinion Public opinion as a concept gained credence with the rise of 'public' in the eighteenth century.
P3M3	P3M3, Programme and Project Management Maturity Model is a reference guide for structured best practice. It breaks down the broad disciplines of portfolio, programme and project management into a hierarchy of Key Process Areas (KPAs). The hierarchical approach enables organisations to assess their current capability and then plot a roadmap for improvement prioritised by those KPAs which will make the biggest impact on performance.
Financial Services	Financial services are the economic services provided by the finance industry, which encompasses a broad range of organizations that manage money, including credit unions, banks, credit card companies, insurance companies, consumer finance companies, stock brokerages, investment funds and some government sponsored enterprises. As of 2004, the financial services industry represented 20% of the market capitalization of the S&P 500 in the United States. History of financial services The term 'financial services' became more prevalent in the United States partly as a result of the Gramm-Leach-Bliley Act of the late 1990s, which enabled different types of companies operating in the U.S. financial services industry at that time to merge.
Personal branding	Personal branding is, for some people, a description of the process whereby people and their careers are marked as brands. It has been noted that while previous self-help management techniques were about self-improvement, the personal branding concept suggests instead that success comes from self-packaging. Further defined as the creation of an asset that pertains to a particular person or individual; this includes but is not limited to the body, clothing, appearance and knowledge contained within, leading to an indelible impression that is uniquely distinguishable.
Storyboard	Storyboards are graphic organizers in the form of illustrations or images displayed in sequence for the purpose of pre-visualizing a motion picture, animation, motion graphic or interactive media sequence.

The storyboarding process, in the form it is known today, was developed at the Walt Disney Studio during the early 1930s, after several years of similar processes being in use at Walt Disney and other animation studios. Origins

The storyboarding process can be very time-consuming and intricate.

Podcasts	A podcast is a type of digital media consisting of an episodic series of audio radio, video, PDF, or ePub files subscribed to and downloaded through web syndication or streamed online to a computer or mobile device. The word is a neologism derived from 'broadcast' and 'pod' from the success of the iPod, as podcasts are often listened to on portable media players. In the context of Apple devices, the term 'Podcasts' refers to the audio and video version of podcasts, whereas the textual version of podcasts are classified under the app known as Newsstand.
Work experience	Work experience is the experience that a person has been working, or worked in a specific field or occupation. The phrase is sometimes used to mean a type of volunteer work that is commonly intended for young people -- often students -- to get a feel for professional working environments. This usage is common in the United Kingdom, while the American equivalent is intern.
Appearance	In law, appearance is the coming into court of either of the parties to a lawsuit, and/or the formal act by which a defendant submits himself to the jurisdiction of the court. Legal details (outdated) The defendant in an action in the High Court of England enters his appearance to the writ of summons by delivering, either at the central office of the Supreme Court, or a district registry, a written memorandum either giving his solicitor's name or stating that he defends in person. He must also give notice to the plaintiff of his appearance, which ought, according to the time limited by the writ, to be within eight days after service; a defendant may, however, appear any time before judgment.
Request for production	A request for production is a legal request for documents, electronically stored information, or other tangible items. In civil procedure, during the discovery phase of litigation, a party to a lawsuit may request that another party provide any documents that it has that pertain to the subject matter of the lawsuit. For example, a party in a court case may obtain copies of e-mail messages sent by employees of the opposing party.
Production	In economics, production is the act of creating output, a good or service which has value and contributes to the utility of individuals. The act may or may not include factors of production other than labor.

Chapter 18. Building Careers and Writing Résumés

Text file	A text file is a kind of computer file that is structured as a sequence of lines of electronic text. A text file exists within a computer file system. The end of a text file is often denoted by placing one or more special characters, known as an end-of-file marker, after the last line in a text file.

Content	In mathematics, a content is a real function μ defined on a field of sets $\mathcal{A}$ such that $\mu(A) \in [0, \infty]$ whenever $A \in \mathcal{A}$. $\mu(\varnothing) = 0$. $$\mu(A_1 \cup A_2) = \mu(A_1) + \mu(A_2) \text{ whenever } A_1, A_2 \in \mathcal{A} \text{ and } A_1 \cap A_2 = \varnothing.$$ A very important type of content is a measure, which is a σ-additive content defined on a σ-field. Every measure is a content, but not vice-versa.

1. _____ and career coaching are similar in nature to traditional counseling. However, the focus is generally on issues such as career exploration, career change, personal career development and other career related issues. Typically when people come for _____ they know exactly what they want to get out of the process, but are unsure about how it may work.

   a. Jimmy Carter
   b. Multifactor
   c. Simple interest
   d. Career counseling

2. _____s are graphic organizers in the form of illustrations or images displayed in sequence for the purpose of pre-visualizing a motion picture, animation, motion graphic or interactive media sequence.

   The storyboarding process, in the form it is known today, was developed at the Walt Disney Studio during the early 1930s, after several years of similar processes being in use at Walt Disney and other animation studios. Origins

   The storyboarding process can be very time-consuming and intricate.

   a. comic strip
   b. Jimmy Carter
   c. Multifactor
   d. Storyboard

3. . A _____ (also spelled curriculum vitæ), provides an overview of a person's experience and other qualifications.

In some countries, a CV is typically the first item that a potential employer encounters regarding the job seeker and is typically used to screen applicants, often followed by an interview, when seeking employment.

_____ is a Latin expression which can be loosely translated as [the] course of [my] life.

a. Delivery order
b. Memorandum of association
c. Curriculum vitae
d. Sales order

4. A business _____ is a written offer from a seller to a prospective buyer. Business _____s are often a key step in the complex sales process--i.e., whenever a buyer considers more than price in a purchase.

There are three distinct categories of business _____s:•formally solicited•informally solicited•unsolicited.

Solicited _____s are written in response to published requirements, contained in a Request for _____ (RFP), Request for Quotation (RFQ), Request for Information (RFI) or an Invitation For Bid (IFB).

a. Qualified prospect
b. Quosal
c. Proposal
d. Sale and rent back

5.

_____ is generally considered an altruistic activity, intended to promote good or improve human quality of life. It is considered as serving the society through own interest, personal skills or learning, which in return produces a feeling of self-worth and respect, instead of money. _____ is also famous for the skill development, to socialize and to have fun.

a. Jimmy Carter
b. Multifactor
c. Volunteering
d. Sale and rent back

1. d
2. d
3. c
4. c
5. c

You can take the complete Chapter Practice Test

for Chapter 18. Building Careers and Writing Résumés
on all key terms, persons, places, and concepts.

Online 99 Cents

http://www.epub2.4.21191.18.cram101.com/

Use www.Cram101.com for all your study needs

including Cram101's online interactive problem solving labs in

chemistry, statistics, mathematics, and more.

Chapter 19. Applying and Interviewing for Employment

_____ | Social network

_____ | Proposal

_____ | Public speaking

_____ | Bureau of Labor Statistics

_____ | Action item

_____ | Job interview

_____ | Storyboard

_____ | Structured interview

_____ | Workforce management

_____ | Job fair

_____ | Viral marketing

_____ | Online interview

_____ | Personality test

_____ | Substance abuse

_____ | Cultural identity

_____ | Negotiation

_____ | Self-consciousness

_____ | Equal Employment Opportunity Commission

_____ | Appearance

	Nonverbal communication
	Extension
	Nonverbal
	Resignation
	Business communication

CHAPTER HIGHLIGHTS & NOTES: KEY TERMS, PEOPLE, PLACES, CONCEPTS

Social network	A social network is a social structure made up of a set of actors (such as individuals or organizations) and the dyadic ties between these actors. The social network perspective provides a clear way of analyzing the structure of whole social entities. The study of these structures uses social network analysis to identify local and global patterns, locate influential entities, and examine network dynamics.
Proposal	A business proposal is a written offer from a seller to a prospective buyer. Business proposals are often a key step in the complex sales process--i.e., whenever a buyer considers more than price in a purchase.

There are three distinct categories of business proposals:•formally solicited•informally solicited•unsolicited.

Solicited proposals are written in response to published requirements, contained in a Request for Proposal (RFP), Request for Quotation (RFQ), Request for Information (RFI) or an Invitation For Bid (IFB). |
| Public speaking | Public speaking is the process of speaking to a group of people in a structured, deliberate manner intended to inform, influence, or entertain the listeners. It is closely allied to 'presenting', although the latter has more of a commercial advertisement. |

Chapter 19. Applying and Interviewing for Employment

Bureau of Labor Statistics	The Bureau of Labor Statistics is a unit of the United States Department of Labor. It is the principal fact-finding agency for the U.S. government in the broad field of labor economics and statistics. The BLS is a governmental statistical agency that collects, processes, analyzes, and disseminates essential statistical data to the American public, the U.S. Congress, other Federal agencies, State and local governments, business, and labor representatives. The Bureau of Labor Statistics also serves as a statistical resource to the Department of Labor.
Action item	In management, an action item is a documented event, task, activity, or action that needs to take place. Action items are discrete units that can be handled by a single person. Action items are usually created during a discussion by a group of people who are meeting about one or more topics and during the discussion it is discovered that some kind of action is needed.
Job interview	A job interview is a process in which a potential employee is evaluated by an employer for prospective employment in their company, organization, or firm. During this process, the employer hopes to determine whether or not the applicant is suitable for the role. A job interview typically precedes the hiring decision, and is used to evaluate the candidate.
Storyboard	Storyboards are graphic organizers in the form of illustrations or images displayed in sequence for the purpose of pre-visualizing a motion picture, animation, motion graphic or interactive media sequence. The storyboarding process, in the form it is known today, was developed at the Walt Disney Studio during the early 1930s, after several years of similar processes being in use at Walt Disney and other animation studios. Origins The storyboarding process can be very time-consuming and intricate.
Structured interview	A structured interview is a quantitative research method commonly employed in survey research. The aim of this approach is to ensure that each interview is presented with exactly the same questions in the same order. This ensures that answers can be reliably aggregated and that comparisons can be made with confidence between sample subgroups or between different survey periods.
Workforce management	Workforce management encompasses all the activities needed to maintain a productive workforce. Under the umbrella of human resource management, WFM is sometimes referred to as HRMS systems, or even part of ERP systems. Recently, the concept of workforce management has begun to evolve into workforce optimisation.

Job fair	A job fair is also referred commonly as a career fair or career expo. It is a fair or exposition for employers, recruiters and schools to meet with prospective job seekers. Expos usually include company or organization tables or booths where resumes can be collected and business cards can be exchanged.
Viral marketing	Viral marketing, viral advertising, or marketing buzz are buzzwords referring to marketing techniques that use pre-existing social networks to produce increases in brand awareness or to achieve other marketing objectives (such as product sales) through self-replicating viral processes, analogous to the spread of viruses or computer viruses (cf. internet memes and memetics). It can be delivered by word of mouth or enhanced by the network effects of the Internet.
Online interview	An online interview is a form of online research method. It takes many of the methodological issues raised in traditional face to face or F2F interviews and transfers these online with some key differences. It principally focuses on the conduct of one-to-one exchanges as one-to-many exchanges are usually called online focus groups.
Personality test	A personality test is a questionnaire or other standardized instrument designed to reveal aspects of an individual's character or psychological makeup. The first personality tests were developed in the early 20th century and were intended to ease the process of personnel selection, particularly in the armed forces. Since these early efforts of these test, a wide variety of personality tests have been developed, notably the Myers Briggs Type Indicator (MBTI), the MMPI, and a number of tests based on the Five Factor Model of personality.
Substance abuse	Substance abuse, is a patterned use of a substance (drug) in which the user consumes the substance in amounts or with methods neither approved nor supervised by medical professionals. Substance abuse/drug abuse is not limited to mood-altering or psycho-active drugs. If an activity is performed using the objects against the rules and policies of the matter (as in steroids for performance enhancement in sports), it is also called substance abused.
Cultural identity	Cultural identity is the identity of a group or culture, or of an individual as far as one is influenced by one's belonging to a group or culture. Cultural identity is similar to and has overlaps with, but is not synonymous with, identity politics. Description Various modern cultural studies and social theories have investigated cultural identity.
Negotiation	Negotiation is a dialogue between two or more people or parties, intended to reach an understanding, resolve point of difference, or gain advantage in outcome of dialogue, to produce an agreement upon courses of action, to bargain for individual or collective advantage, to craft outcomes to satisfy various interests of two people/parties involved in negotiation process.

	Negotiation is a process where each party involved in negotiating tries to gain an advantage for themselves by the end of the process. Negotiation is intended to aim at compromise.
Self-consciousness	Self-consciousness is an acute sense of self-awareness. It is a preoccupation with oneself, as opposed to the philosophical state of self-awareness, which is the awareness that one exists as an individual being; although some writers use both terms interchangeably or synonymously. An unpleasant feeling of self-consciousness may occur when one realizes that one is being watched or observed, the feeling that 'everyone is looking' at oneself.
Equal Employment Opportunity Commission	The U.S. Equal Employment Opportunity Commission is a federal law enforcement agency that enforces laws against workplace discrimination. The Equal Employment Opportunity Commission investigates discrimination complaints based on an individual's race, color, national origin, religion, sex, age, disability, genetic information and retaliation for reporting, participating in and/or opposing a discriminatory practice. The Commission also mediates and settles thousands of discrimination complaints each year prior to their investigation.
Appearance	In law, appearance is the coming into court of either of the parties to a lawsuit, and/or the formal act by which a defendant submits himself to the jurisdiction of the court. Legal details (outdated) The defendant in an action in the High Court of England enters his appearance to the writ of summons by delivering, either at the central office of the Supreme Court, or a district registry, a written memorandum either giving his solicitor's name or stating that he defends in person. He must also give notice to the plaintiff of his appearance, which ought, according to the time limited by the writ, to be within eight days after service; a defendant may, however, appear any time before judgment.
Nonverbal communication	Nonverbal communication is usually understood as the process of communication through sending and receiving wordless (mostly visual) between people. Messages can be communicated through gestures and touch, by body language or posture, by facial expression and eye contact. Nonverbal messages could also be communicated through material exponential; meaning, objects or artifacts (such as clothing, hairstyles or architecture).
Extension	In any of several studies that treat the use of signs - for example, in linguistics, logic, mathematics, semantics, and semiotics - the extension of a concept, idea, or sign consists of the things to which it applies, in contrast with its comprehension or intension, which consists very roughly of the ideas, properties, or corresponding signs that are implied or suggested by the concept in question. In philosophical semantics or the philosophy of language, the 'extension' of a concept or expression is the set of things it extends to, or applies to, if it is the sort of concept or expression that a single object by itself can satisfy.

Nonverbal	Nonverbal communication is usually understood as the process of communication through sending and receiving wordless (mostly visual) cues between people. Messages can be communicated through gestures and touch, by body language or posture, by facial expression and eye contact, which are all considered types of nonverbal communication. Speech contains nonverbal elements known as paralanguage, including voice quality, rate, pitch, volume, and speaking style, as well prosodic features such as rhythm, intonation, and stress.
Resignation	A resignation is the formal act of giving up or quitting one's office or position. It can also refer to the act of admitting defeat in a game like chess, indicated by the resigning player declaring 'I resign', turning his king on its side, extending his hand, or stopping the chess clock. A resignation can occur when a person holding a position gained by election or appointment steps down, but leaving a position upon the expiration of a term is not considered resignation.
Business communication	Business Communication: communication used to promote a product, service, or organization; relay information within the business; or deal with legal and similar issues. It is also a means of relaying between a supply chain, for example the consumer and manufacturer. Business Communication is known simply as 'communications'.

1. _____ communication is usually understood as the process of communication through sending and receiving wordless (mostly visual) cues between people.

   Messages can be communicated through gestures and touch, by body language or posture, by facial expression and eye contact, which are all considered types of _____ communication. Speech contains _____ elements known as paralanguage, including voice quality, rate, pitch, volume, and speaking style, as well prosodic features such as rhythm, intonation, and stress.

   a. Jimmy Carter
   b. On the fly
   c. Nonverbal
   d. Appropriation

2. . A _____ is a social structure made up of a set of actors (such as individuals or organizations) and the dyadic ties between these actors. The _____ perspective provides a clear way of analyzing the structure of whole social entities.

Chapter 19. Applying and Interviewing for Employment

The study of these structures uses _____ analysis to identify local and global patterns, locate influential entities, and examine network dynamics.

a. Social network
b. Multifactor
c. Simple interest
d. Gramm-Leach-Bliley Act

3. _____s are graphic organizers in the form of illustrations or images displayed in sequence for the purpose of pre-visualizing a motion picture, animation, motion graphic or interactive media sequence.

The storyboarding process, in the form it is known today, was developed at the Walt Disney Studio during the early 1930s, after several years of similar processes being in use at Walt Disney and other animation studios. Origins

The storyboarding process can be very time-consuming and intricate.

a. comic strip
b. Jimmy Carter
c. Storyboard
d. Agile manufacturing

4. A _____ is a process in which a potential employee is evaluated by an employer for prospective employment in their company, organization, or firm. During this process, the employer hopes to determine whether or not the applicant is suitable for the role.

A _____ typically precedes the hiring decision, and is used to evaluate the candidate.

a. Jimmy Carter
b. Administratium
c. Adversarial purchasing philosophy
d. Job interview

5. . A business _____ is a written offer from a seller to a prospective buyer. Business _____s are often a key step in the complex sales process--i.e., whenever a buyer considers more than price in a purchase.

There are three distinct categories of business _____s:•formally solicited•informally solicited•unsolicited.

Solicited _____s are written in response to published requirements, contained in a Request for _____ (RFP), Request for Quotation (RFQ), Request for Information (RFI) or an Invitation For Bid (IFB).

a. Proposal
b. Quosal
c. Request for proposal

1. c

2. a

3. c

4. d

5. a

You can take the complete Chapter Practice Test

for Chapter 19. Applying and Interviewing for Employment
on all key terms, persons, places, and concepts.

Online 99 Cents

http://www.epub2.4.21191.19.cram101.com/

Use www.Cram101.com for all your study needs

including Cram101's online interactive problem solving labs in

chemistry, statistics, mathematics, and more.

Other Cram101 e-Books and Tests

Want More?
Cram101.com...

Cram101.com provides the outlines and highlights of your
textbooks, just like this e-StudyGuide, but also gives you the
PRACTICE TESTS, and other exclusive study tools for all of your
textbooks.

Learn More. *Just click*
http://www.cram101.com/

Lightning Source UK Ltd
Milton Keynes UK
UKOW06f1459200913

217589UK00001B/30/P